Political Ecology of Everyday Resistance and State Building

Resource extraction and conflicts over natural resources are a global phenomenon, including India. Indigenous tribes, like the Ho community in Jharkhand, are affected by these dynamics, as their cultural practices and livelihoods are intertwined with the local ecology.

This book explores the process of State formation through developmental intervention in the resource-rich areas of Jharkhand in Eastern India which are inhabited by the indigenous Ho community.

The conflict in Jharkhand is intertwined with State development projects and capitalist interventions. This book examines the history of these projects and the issues of territorialization, dispossession, accumulation, and marginalization which the communities have been fighting against for many decades. It examines the process of development policies and projects shaping and restructuring the resource-rich ecology of the region and addresses the interrelated issues of development-induced dispossession, resistance, ecological transformation, governance, illegalities, and State building. It focuses on the following questions: What do development projects bring to the Ho community? What induces them to resist and negotiate? How do State decentralization schemes and local governance in resource-conflict areas strengthen State capacities?

The book highlights the consequences of the livelihoods and cultural practices of the local people because of ecological transformation and everyday resistance.

Comprehensive and important, this book will be of interest to students and researchers of anthropology, sociology, political ecology, social work, development studies, ecology, developmental sociology, indigenous studies, law, and economic anthropology.

Dhiraj Kumar is an assistant professor of Sociology at the Mahila Maha Vidhyalaya, Banaras Hindu University, India.

Political Ecology of Everyday Resistance and State Building

A Case of the Ho of Jharkhand

Dhiraj Kumar

Routledge
Taylor & Francis Group
LONDON AND NEW YORK

First published 2024
by Routledge
4 Park Square, Milton Park, Abingdon, Oxon OX14 4RN

and by Routledge
605 Third Avenue, New York, NY 10158

Routledge is an imprint of the Taylor & Francis Group, an informa business

British Library Cataloguing-in-Publication Data
A catalogue record for this book is available from the British Library

ISBN: 978-1-032-34570-3 (hbk)
ISBN: 978-1-032-63801-0 (pbk)
ISBN: 978-1-032-63799-0 (ebk)

DOI: 10.4324/9781032637990

Typeset in Sabon
by SPi Technologies India Pvt Ltd (Straive)

To Baba, Papa and Maa

Contents

Illustrations

Tables

Map

Foreword

In India, terms like 'Adivasis' and 'Indigenous people' are often used interchangeably to refer to the scheduled tribes, although they carry distinct connotations and implications depending on the context in which they are analyzed. Government of India in its FRA, 2006 has recognized that these people were subjected to historical injustice for centuries that need to be rectified. This act recognizes the rights of the forest dwelling tribal communities and other traditional forest dwellers to forest resources on which these communities are dependent for a variety of needs, such as livelihood and habitation and many other sociocultural ones. This act of the constitution promised to ensure land tenure, livelihood, and food security and authorized the forest right holders to preserve resources through local self-government (Gram Sabha) for sustainable use, conservation of biodiversity, and maintenance of ecological balance. The post-colonial State authorities have been claiming that numerous 'development' interventions have been made in the Adivasi regions. However, these contributed to further sharpening of social inequalities. Thus, the growth capital at the cost of the livelihood of the people in the newly formed mineral-rich state Jharkhand manifested the tenets of political ecology with its increasing intersectional inequality between the local ecology and global political for economic processes.

Since early 1990s, liberalization has generated a global capital – a dominant form of economy across the world that is no more restricted within the boundary of any nation State. Many multinational players for business have entered into non-renewal extractive industries, such as the mining sector in developing nations wherein the State assumes the role of the acquiring body to meet the need of the requiring bodies – the corporates. It is an accepted fact that capitalists in India keep close alliance with the undemocratic elements that take the mass to a make-believe world. Such a process can hardly have the capacity to generate mass employment for the poor, neither does it have the capability to reduce the gap between the rich and the poor. Technological automation in mining operations is an anti-dose for labour-intensive employment. The Mines and Minerals (Development and Regulation) Amendment Act, 2015 has ignored the need of consent of Gram Sabha.

Thus, it wilfully ignored the people's perspective of development. Dr Dhiraj, in his study, focused on the impact of development projects and policies of the State on Adivasi's ecosystem and socio-economic frontiers. The hitherto indigenous republics in the Adivasi region have been engulfed by national political system and gradually co-opted into the modern State system to remain at the receiving ends. In the process, the resource-rich Adivasi habitat slides down towards pauperization. There has been a long interest of conflict between the State and the local actors over the control, use, and management of natural resources.

The wrath of mineral-extracting corporates and State accumulative projects exponentially divested the Adivasi habitat, even passing through societal sickness of State–community conflicts, resistance, and arbitration. The use of forest for no forest purposes deprived the Adivasis of their customary rights and livelihood resources. Dr Dhiraj meticulously has explained the interrelated issues of development-induced dispossession, resistance, ecological transformation, and tribal resistance against the extractive projects. Here, positioning refers to the position of the Adivasis as a distinct category of being. Particularly, this book makes a critical assessment of the State–Adivasi relationships.

The work is extremely impressive from the viewpoint of theoretical arguments as well as the description of the struggling lives and strangled voices of the natives. The book has attempted to surface several abstruse facts in a diaphanous manner as well as posing several political questions of critical angularities. I am sure that this book will be immensely helpful to development planners, sensible researchers, executives, activists, academicians, NGO personnel, and implementing agencies for welfare measures.

Prof. Premananda Panda
Former Professor,
Department of Anthropology,
Sambalpur University
and
Former Director,
SC, ST and RTI, Government of Odisha

Preface and Acknowledgements

I would like to express my sincere gratitude to several individuals who have supported and encouraged me during the course of this book. Firstly, I extend my special thanks to Professor Premananda Panda, Professor R Siva Prasad, and Professor Lancy Lobo, Dr. N. R. Mishra and Amit Kumar, for their valuable suggestions and encouragement.

I am also grateful to the people at the Johar Office at Chaibasa, West Singhbhum, for their logistical assistance, hospitality, and friendship during my field stay. I extend my heartfelt thanks to all the key informants and people for their human compassion and support during this time. I would also like to acknowledge my senior friend Dhananjay Kumar for his encouragement and timely discussions. I also extend my thanks to Dr. Dinabandhu, Dr. Gautam, Mr. Sarang Hota, and other friends for their emotional support.

I extend my deepest appreciation to my father, mother, sister, brothers (Niraj and Pankaj), and wife Kalpana for their unwavering support, encouragement, and helpfulness. I cannot express enough gratitude for the smile and emotion of my adorable son Dakshesh, who has stimulated positive energy in my veins. I would also like to thank Prof. Rita Singh and my fellow colleagues at MMV, BHU, for their support.

Last but not least, my biggest appreciation goes to all my key informants in the field for sharing their experiences and providing their valuable time. I could not have accomplished this without their sincere openness and cooperation, and I am truly grateful for getting to know all of them. It has been a pleasure to work with Shoma Choudhury and her team at Routledge. I acknowledged Suba Ramya and the entire team for their support in production of this book.

Finally, I would like to conclude this list of acknowledgements by quoting Tim Ingold, who reminds us that intellectual sources are not the only things that move us but also the many natural elements and companions we share our lives with. If our aim is to read the world, as I believe it should be, then the purpose of written texts should be to enrich our reading so that we might be better advised by, and be responsive to, what the world is telling us.

Abbreviations

ACC	Associated Cement Companies
AJSU	All Jharkhand Student Union
ASHA	Accredited Social Health Activist
ATMA	Agricultural Technology Management Agency
CSR	Corporate Social Responsibility
CPI	Communist Party of India
CNT	Chotanagpur Tenancy Act
DI	Development Intervention
GDP	Gross Domestic Product
GVS	Gram Vikas Samitis
JPRA	Jharkhand Panchayati Raj Act
JMM	Jharkhand Mukti Morcha
KES	Kolhan Estate System
KGE	Kolhan Government Estate
MNREGA	Mahatma Gandhi National Rural Employment Guarantee Act
MLA	Member of legislative Assembly
MMDR	Mines and Minerals (Development and Regulation) Act
MP	Member of Parliament
MOU	Memorandum of Understanding
PCCF	Principle Chief Conservation of Forest
PIL	Public Interest Litigation
PESA	The Provisions of the Panchayats (Extension to Scheduled Areas) Act
PMAY	Pradhan Mantri Awas Yojna
ST	Scheduled Tribe
SWFA	South West Frontier Area
VLW	Village-Level Worker

Glossary

Bera land	Bera land was fertile agricultural land used for *aghani* (winter rice) paddy, and it was safe from drought as it was irrigated from the river streams, tanks, and springs, whereas bad land required a good rainfall.
Bongas	Spirits
Dikus	External to the tribal region
Diyang	Rice beer
Gora and Bakai	*Gora* is roughly cultivated, whereas *bakai* or *bari* were cultivated, and two crops, usually corn and mustard, were cultivated
Haat	Weekly market
Hatu	Village
Khuntkattidar	Original settlers are known as *Khuntakattidar*
Killi	Clan
Kolhan	Kolhan, the land of the Kol or Ho community.
Manki	
Manki Munda	traditional tribal governance system
Mundas	Customary village chief
Nazarana	The form of cash and livestock
Parha Raja	Customary institution of tribal community
Pashchimi	West Singhbhum, popularly known as Pashchimi
Singhbhum	Singhbhum, is one of the 24 districts of Jharkhand. This region is also known for the Manki Munda system of traditional governance
Sal	Shorea robusta
Sasandiri	Burial ground
Singhbonga	Tribal Sun God, the supreme deity who created the universe
Tola	Hamlet or small village

Chapter 1

State and Adivasi

Reappraising Development, Resistance, and State Formation

> SAIL and Rungta get a reserve forest to mine in West Singhbhum.[1]
> 2000 hectares of the forest have been diverted for mining under the forest conservation act by the environmental ministry.[2]
>
> On 26 September 2012, Ho Adivasi of West Singhbhum opposed the construction of the ICHA dam, which is a multipurpose project on Subernarekha Dam. The protest is continuing to date; the World Bank also withdrew the funding due to opposition and conflict between State and Adivasi.[3]
>
> 31 August 2014, Justice M. B. Shah reported illegal mining in the region of West Singhbhum. The report stated that Iron ore was extracted illegally from 1993 to 2013 without environmental clearance.[4]

What connects these different events is not only the conflict and struggle of Adivasi with the other actors to defend their ecology, which gave the physical, social, and cultural bases upon which the lives and livelihoods of the Adivasi depend. These events also show ecological disequilibrium in the region due to the rise of the commodity frontier and the expansion of State intervention. Contemporary India observes such environmental conflicts and the conflicts between local communities and state-backed industries in resource-rich areas. The roots of the conflicts are embedded in the discussion of the form and character of development that have been facilitated by the State to boost corporate capitalism (Prasad, 2016). Most of the populations that resist the State extractive projects and schemes are the Adivasis and the peasants. Here, it is essential to mention that 'Adivasis', 'Tribes', and 'Indigenous' are three different terms that are related to each other. The term 'Adivasi' essentially refers to a form of self-expression that has arisen in opposition to the State-led process of development-induced dispossession.

Additionally, it represents a political identity of the protest whose alternate means of production, imagination, and polity differ from the mainstream.

The social relations and ecological system determine Adivasi protests and conflicts. Their lives and worlds are culturally connected with the surrounding ecology (Padel & Das, 2010; Prasad, 2008). They often use the 'phrase of the

DOI: 10.4324/9781032637990-1

place' to describe not merely their physical location but to convey how deeply they are culturally connected with their ecology (Wildcat & Pierotti, 2000). It provides them with a sense of identity, unity, and resistance. In contrast, tribe primarily refers to a State-recognized identity that has existed after the State's formulation of the hegemonic paternalistic role, which structured the class- and community-based mobilizations through State-driven institutionalized identity (Prasad, 2016; Ratnagar, 2003). Scholars like Hardiman (1987) advocate for the acceptance of the term 'Adivasis' for the ecological natives of India.

Interestingly, Chotanagpur is generally known as the birthplace of the term 'Adivasi'. Still, the usage of the term Adivasi is absent in northeastern India, where 11 percent of India's so-called tribal population lives (Xaxa, 2016). People of this region prefer to call themselves tribes and use the term 'Adivasis' to refer to those who migrated from other areas during the colonial period to work in tea plantations (Xaxa, 2008, 2016). To avoid delving into the complexities of epistemology and ontology, it is important to acknowledge that the terms 'Tribe', 'Adivasi', and 'Indigenous' are often used interchangeably, yet they carry distinct connotations and implications depending on the context in which they are employed. In general, these terms refer to groups of people who have lived in a particular region for generations, have their own distinct culture, and may have experienced historical marginalization and discrimination at the hands of dominant societies. However, the specific nuances and meanings of each term can vary depending on factors such as cultural and political contexts, historical relationships between different groups of people, and the perspectives and intentions of the individuals using the terms.[5]

Adivasi communities are intrinsically linked to their environment through different strategies, that is economic exchange, ceremonies, language, and spiritual tradition. In this context, ecology has become essential in disputes involving transnational corporations, governments, and Adivasi communities. Thus, ecology plays a pivotal role in power relations between different stakeholders. According to Giddens (1979), all power relations are two-way relations of autonomy and dependence, regardless of an unbalanced distribution of resources. The unstable distributions of resources are intrinsic to the State's formation and its development policy (Bryant, 1998). State formation or State-making refers to a process that shaped the forms and legitimations of government and governmentality (Sivaramakrishnan, 1996). It is the analysis of pathways and modes through which locality, region, and transnational bodies became spatial and cultural entities. The pre-State indigenous republics in the Adivasi region gradually became a part of the modern State system and formation.[6] However, Adivasis of resource-rich regions are facing a phase of State accumulative projects at an unprecedented rate. It sets the theatre of conflicts, resistance, and negotiation. These can be termed 'socio-environmental conflicts' (Muradian, Walter, & Alier, 2016).

The expansion of the capitalist forces in the name of development projects (dam, mining, and diversion of forest for extractive industry) subjugated the Adivasi ecosystem. It is a fact that the issues of land and forest rights of Adivasis, especially in the resource-rich areas in India, have been highly subtle because they have been deprived of their customary rights for purposes such as large dams, mining, timber contracts, or biodiversity conservation by the State. Numerous development interventions have been taken in the Adivasi regions, but they have sharpened social inequalities (Xaxa, 2011).

The mineral extraction path and development intervention have led to a growing number of struggles in which access by socially disadvantaged groups (Bryant, 1992) to the environment that constitutes the base of their livelihoods and culture is the dependent threat from the intervention of influential actors. The powerful actors have different means of resource use than traditional actors, especially Adivasis. Hence, the resistance and struggle show the power relations that shape access to natural resources. Power asymmetry is concerned with the everyday manner in which access and conflict are negotiated, deflected, or pre-empted. The issues of conflict and access to the natural resource base are not new phenomena. Conflicts over the control and use of natural resources and their recognition have been taking place in the political sphere. This is because the State and the Adivasi community are interested in the same resource base. There has been a long conflict of interest between the State and the local actors over the control, use, and management of natural resources. The following are the issues that constantly emerge: the role of the State in managing the resource and the politics of exclusion and inclusion; environmental justice; participation, democracy, and power equality; right; and entitlement, particularly for the local communities who rely mainly on the land and forest. Recently, the scope and magnitude of conflict over natural resources have increased and intensified. These developmental projects can heighten resistance movements and cause environmental degradation, disruption of projects, and economic change in the mode of resource use. Analyzing the relationship between domination and change requires continuous engagement with historical and contemporary periods. So, there is a need to understand the State–Adivasi interaction in the context of sociopolitical structure that influences the local ecosystem, which Vayda (1983) terms 'progressive contextualization'.[7] This study focuses on how development projects and policies of the State impact and shape Adivasis' ecology and economics. It concerns the State's 'will to develop Adivasi' in the representative form of State formation and developmental activities. The new interest in the State arises from the recognition of the central role that the State plays: a dominating position in shaping the ecology and economics of 'local communities that have historically constituted the objects of anthropological and sociological inquiry.

Here, a discussion of State is extremely crucial in this context. As (Whitehead, Jones, & Jones, 2007, p. 14) argue:

> we are suspicious of discussions of states interacting with natures ... because such phrases tend to suggest two already established entities doing things to each other – the state affecting the natural environment, or changes in nature somehow influencing the policies and dictates of the state ... in order to analyse state-nature relations it is important to unpack both categories so as to understand how their constituent parts are intermixed and connected.
>
> (Whitehead, Jones, & Jones, 2007)

This study addresses the interrelated issues of development-induced dispossession, resistance, ecological transformation, tribal resistance against the extractive projects, shadowy practices of democracy, illegalities, and State building to unpack the State role in shaping the ecology and tribal lifeworlds. In this study, I try to understand the tribal's critique of State developmentalism and conflict over development to explore the 'positioning' (Li 2000). Here, positioning refers to the position of Adivasis as a distinct category of being. Particularly, this book makes a critical assessment of the State–Adivasi relationships.

Tribals of India have had a long history of struggle and conflict with the colonial and post-colonial State. Their struggle can be conceptualized as ethnic or environmental conflict (Jewitt, 2008) against the expansion of accumulative projects. The discourse and movement against State developmentalism were related to ethno-conflicts and livelihood issues, but the adaptation of industrialization and resource extraction as the primary chosen path of development has accelerated accumulative and resistance activities. The predominately natural resource region inhabited by Adivasis started to experience a substantial expansion of the region-specific commodity frontier in the name of development projects and schemes that also disturbed the local ecosystem in a velocitized manner. For mining of natural resources, more than 2.2 lakh hectares of land have been given out on lease (George, 2009) ignoring the customary and constitutional rights of the Adivasis over the local ecology and natural resources in Jharkhand only. The Adivasi societies of Jharkhand have been studied mainly in terms of their culture and economy (Hebber, 2011; Nathan & Kelker, 2003; Prakash, 2011; Shah, 2013), but the focus on State accumulative and shadowy character from the political ecology perspective has got less attention. Several studies (Prakash, 1999; Sachchidananda, 1972; Weiner, 1978) have argued that development projects have led to dissatisfaction among the Adivasis of this region. Industrialization, land alienation, and migration from the Adivasi areas are the main reasons for the change. This study also talks about how legal pluralism of the State in the resource-rich region of Adivasi favours the capitalist class

(George, 2009; Sundar, 2005) and treats the natural resource-rich region as a colony (Puthumattathil, 2014). Additionally, Shah's (2010) seminal work on the Munda tribe showed how the Munda experienced State as an alien subject and challenged State legitimacy through sacral and customary practices. Verardo (2003) talked about the de-Sanskritization process to deal with the social and political transformations of the Ho and Munda tribes. Indeed, over the years, a substantial body of knowledge concerning the development and its impact on tribal culture has been built up, but from the political ecology perspective, very few studies have been undertaken that deal with everyday State practices and building programmes in resource-conflict areas where Adivasis resist and negotiate with the State at the agency level through their customary institutions. The unanswered questions of this field of inquiry are: (a) How does the State as an actor interact with Adivasis and other actors in a region-specific trajectory? (b) What is the politics of access, and to what extent is development meant for tribals? (c) What are the structural changes due to resource-conflict and State-led development? (d) What aggravates conflict, resistance, and negotiations? (e) How do States nurture participatory governance through shadowy practices of democracy and how do they co-opt the local institutions? These are some important questions set out to analyze how the State stimulates the extractive players in resource-rich regions of Adivasis which induces power asymmetry and trajectories of injustice and assimilation. To do so, this study specifies the region of West Singhbhum, a Ho-dominated territory, as a research base area and a concerned area within the Jharkhand state to discuss the process of State formation. It also addresses the interrelated issues of development-induced dispossession, marginalization, resistance, ecological transformation, illegalities, co-optation, and assimilation.

1.1 The Adivasis and the State

Adivasis are State-fleeing and challenging people (Saha, 1986). They have a powerful, distinct sociocultural group and organization with alternative modes of production and practices (Sinha, 1993; Skaria, 2001) that differ from the mainstream mode of production. Kin-ordered mode, weak material foundation, inability to produce a surplus, democratic value, and technological lag were the features of Adivasi polity (Sen, 2011, 2018). In the case of the Munda community of Jharkhand, Shah (2010) provides some fascinating details about their alternative visions, radical democratic values through Parha Raja (customary institution), and alternative visions of development and the State. Adivasi cultural complex is based on magico-religious beliefs – primary civilizations, although from a cultural evolutionist perspective, that provide us with their self-restrictive and egalitarian, interaction with other beings, including all living and non-livings. This is an antidote to neoliberal capitalist regimes. The tribal notions of humans stand in sharp contrast to the Western anthropocentric-turned-ethnocentric-turned-technocratic theory of humans (Saraswati, 1991).

From a historical point of view, they are a group of people without history, and their written record has been solely connected to State-making. They are non-State people, and their pasts are known only in relation to State-building projects (Zou, 2016, p. 98). Discussion on ethnicity and tribe begins exactly where taxes and sovereignty of State end (Scott, 2010).

For the Adivasis, cultural practices should be seen as a mode of production for the sustainable use of natural resources. Historically, they live within and on the fringe of forest areas and mainly depend upon natural resources for their livelihood. Ecology, in general, and its resources, in particular, shape their collective imaginations, belief systems, and social system. This umbilical relationship between tribal people and forests was first officially disturbed in the colonial era and continues in the name of different developmental logic. The industrial revolution and the expansion of infrastructural development stimulated the demand for timber, which motivated India's colonial government to govern natural resources and forests through policies and laws (Damodaran, 1997; DasGupta, 2009; Sen, 2011). The introduction of different regimes like forest regime and other forms of bureaucracy were designed for State formation through primitive accumulation. The expansion of State formation and commercial uses of nature were impossible without forcibly taking away the ownership and claims of peasants and tribes regarding their means of production and labour. Many insurgencies have been reported over time. There were nearly more than 110 instances recorded throughout colonial State formation (Guha, 1983).

Adivasi have an ecological culture that must be seen as a part of social relations and forces of production. The political economy of the Adivasis is not economistic, yet; their political economy would be another production paradigm (Padel, Dandekar, & Unni, 2013), forcing them to resist State formation. The category Scheduled Tribe (ST) was created by the colonial administration to counter insurgencies against primitive accumulation, pull insurgent forest dwellers and peasants out of 'natural economies', and bring them within the fold of irreversible universal time for social production and reproduction of capital (Savyasaachi, 2016).

The State regime continues to pull the Adivasi community out of the 'state of nature' and bring them and nature into capitalist production. Construction of the category 'tribe' was used to suppress the community's 'will not to be governed'. Suppression was also followed up with the process of co-optation. Both now determine the relationship between the State and Adivasis in post-colonial India through post-colonial politics of recognition and development (Kapila, 2008). It is essential to see why the acts and development projects of the State which pretend to empower the Adivasis failed to deliver their promise and also why they have designated the State even more than others (Sareen, 2015; Savyasaachi, 2016; Sundar, 2005). Adivasis as a group has gained the least and lost the most from India's six [*seven*] decades of democracy and development (Guha, 2010). In the case of Bhil Adivasis, the State formation produces a culture of fear and deference (Baviskar, 1997b), whereas this kind of society and State relation is called everyday tyranny (Nilsen, 2016). Nilsen refers to it to show how the State imposed a predatory and coercive regime of corruption and violence that the local Bhil community experienced every day.

1.2 Political Ecology and Conflict

Scholars of political ecology do not primarily focus on violent conflicts as they tend to see conflicts and conflicting interests intrinsic to social relations and State–peasant relations. Earlier, they mainly concentrate on human–nature interaction, but in contemporary times, the role of the State in shaping the environment is more than ever (Collier, 2010; Ioris, 2015; Peet & Watts, 2016). Political ecology is the struggle among actors to control the environment and its resources. Resources invoke utility, and conflict indicates a lack of harmony. Therefore, their focus is on the social structures and constructions that shape access to and control natural resources. A natural resource may begin its social life in the State's territorial domain. The social forces of the State and other actors make nature more productive. However, the use and abuse of natural resources is based on the relation of domination (Escobar, 2006b). The description of resources, environment, and conflict are numerous, and their construction is mainly political, whereas the academic discourse is unlikely to follow any single formulation. Understanding conflict is the outcome of interaction of various actors possessing unequal power capabilities. In this vein, Alier (1994) observes that conflict for the livelihood, environment, and ecology can be expressed in terms of the ecological requirement of life. He mentions that ecological requirement is a broad theme embedded in the environment, livelihood, and the political economy of natural resources. In political ecology literature, one can use neo-Malthusian theories of resource scarcity to understand resource conflict. The advocators of this school argue that resource scarcity induced conflicts, and political and economic factors drive it. One of the influential advocators of this school is Homer Dixon. He talks about eco-violence (Gleditsch & Urdal, 2002) by considering resource scarcity as a decisive factor. Hardin refers to the tragedy of commons as the situation when one group or people exploit the common shared resources until the resources become unavailable to some group or all (Hardin, 1968). The unavailability of resources and scarcity of renewable resources induced violence and conflict (Dalby, 2003). Unlike the assumption of neo-Malthusian theories of resource scarcity and conflict, a recent stream of resource-conflict analysis has emerged that argues that the availability of resources and wealth causes conflict (De Soysa & Neumayer, 2007). Advocates of this school (Humphreys, 2005; Le Billon, 2011; Ross, 2004) focus on the political economy of resource use. They argue that the characteristics and values of natural resources determine conflict. They mention that the ways social institutions of the State and its agencies shape the rules and rights of resource uses and access should be the area of inquiry.

Vulnerability results from resource dependence, rather than conventional notions of scarcity or abundance that induce conflict (Le Billon, 2011). Their point is that capitalist greed to maximize the profit unequally distributed the natural resource wealth directed to conflict. 'Greed-motivated' resource conflicts are related to the discussion of rational choice and rent-seeking behaviour (Collier, 2010). Scholars like Omeje (2008) argue that resource conflicts can be viewed as a historical product inseparable from the region's social construction and political economy.

The empirical work of Bassett (1988) on the farmer–herder conflict in the northern Ivory Coast is based on a regional political ecology framework. It argues that examining the role of the State is essential to understand regional variability and conflict. Conflict also occurs due to the expansion of the commodity frontiers that affects the different dimensions of environmental justice (Alier, 1994). A scholar like Moore (1993) employs Gramscian metaphor to analyze conflicts between a differentiated peasantry and the State. His ethnographic study examines the conflicts over access to environmental resources and land in Nyanga District, eastern Zimbabwe. He mainly focuses on the role of State agencies and their ability to influence access to resources. Popular beliefs, historical myths, culture, and ideological differentiation also determine conflict. Struggles over the symbolic meaning and cultural meaning are associated with conflicts over the material relation of production (Moore, 2005). A historian like Grove argues that a new form of laws and policies of the State facilitates conflict. He said:

> It was the imposition of new forms of land designation, as between private and public and in the interruption of customary methodologies of interaction with forest, pasture, and soil that colonial states (and post- colonial states effectively modelled on them) have exercised the most intimate and often oppressive impact on the daily lives and ways of production of the rural majority through much of the (especially tropical) world.
>
> (Grove, 1990, p. 16)

Bryant and Bailey (1997) examine how forest politics in India, Java, and Burma has been conditioned by the struggle of different actors (peasants, shifting cultivators, and businessmen) with the State. They also explained the struggle and conflict within the various agencies of the State (agriculture and forestry) and their impact on the local community. Defence of livelihood against the State expansion or facilitation of capitalist projects also induces conflict (Alier, 1994).

1.3 Political Ecology: Politics of Power Over Environment

Political ecology is not a homogeneous research agenda, as it has vast literature and a variety of critical issues within the politicized nature–society interaction. In this vein, Nightingale (2002) illustrates some significant issues related to politics regarding the environment. She mainly examines the linkage between local ecology and global political-economic processes. She also talks about how capitalist development intervention is linked with the marginalized primary producers on one side and for profit on the other. The role of knowledge is also significant for the issues of claims, access, and transformation of the environment. It is because the different knowledge discourse is used to mobilize nature politically and socially (Escobar, 2006a). This knowledge can be indigenous ecological knowledge and the local communities' strategies or the modern knowledge and strategies of the institution like the State. It means that knowledge is used to promote claims and positions to contest ownership. Understanding discourse and knowledge production is necessary in this regard. Discourse analysis enables political ecologists to plumb the new depth and ask questions such as 'how institutionalised knowledge is privileged and how facts are contested' (Peet & Watts, 2004). Neo-Marxist sense of social scientists uses it to study the power relations that mediate 'human–environment relations'. The term was first used by Wolf (1972) in his study titled Ownership and Political Ecology. Placing power in the centre of analysis, political ecology in this vein differed from cultural ecology and apolitical ecology. Cultural ecology mainly focuses on the problematic adaptation of the environment (Steward, 1972) without discussing the power politics that induces structural inequality and also mediate human relations with ecology (Friedman, 1974; Robbins, 2012). Political ecology combines politics with ecology to provide a chain of explanations to the ecological phenomenon (Blaikie & Brookfield, 1987).

Political ecology is not like the political economy of Marx, in which economic determinism is prioritized over another component of social formation. The ways biophysical ecology is interpreted, contested, and negotiated are highly political (Baginski & Blaikie, 2013). They gave the example of forest ecology to show how this category is socially constructed and contested. Political ecology is a rapidly developing conceptual approach. The two words in political ecology suggest two different meanings. In this vein, ecology is conventionally assumed to be separate from politics as a traditional explanation. This explanation believes ecology is objective, scientific, and separated from culture, whereas political ecology considers that politics and power shape ecology. Ecology is objective, scientific, and rational, but politics is subjective and socially constructed. Different actors shape the ecology with their own beliefs, strategy, and interest. For example, the word 'forest', for a tribal woman, has a different meaning than for others. To tribal women, it is a place of identity, culture, and livelihood, whereas for the State or the strategic planner, it is an administrative category or a site. Agrawal (2005), in his study on the forest of Uttarakhand, explores the interlinkages between power, knowledge, and institutions. He also termed 'environmentality' to denote how the State uses knowledge, power, and politics to govern the environment. Power, politics, and their dynamics shape/reshape the ecology. Schneider's (2011) study on Cambodia examines the everyday politics of peasant resistance against the power dynamics of the land grab. Based on case studies, Schneider tries to understand the role of the State and the character and dynamics of the land grabs.

1.4 Development, Dispossession, and State: Why Political Ecology?

A political ecological perspective, one that foregrounds history, scale, power, and agency, offers a valuable adjunct to understanding development-induced dispossession. Capitalist development and its fallout – dispossession – have been contested by various place-based struggles in India. They have intensified capital accumulation, enforcing the vast majority of the population, particularly the Adivasis in resource-rich territories to displace, and have affected their livelihoods by accumulating their cultural rights to land, water, and forests. The prerequisite capitalist logic of investment-induced dispossession has been contested in various place-based local struggles raising essential questions about mass mobilization, resistance, politics of protest, identity, and solidarity. Contemporary global 'land rush' could be happening at the expense of common property systems worldwide (Dell'Angelo, D'Odorico, Rulli, & Marchand, 2017). They use the term 'common grabbing' to understand the dispossession of the local people. By conducting a meta survey on large-scale land acquisitions at the global level, they argue that common grabbing is an integral component of the land rush, and it preferentially targets the commons. Targeting the commons is equated to the discussion of dispossession that induces movement, resistance, and conflict. The nature of dispossession also shapes the character of anti-dispossession movements. Six fundamental features of any form of politics, that is, target, strategy and tactics, political organization, social composition, goals, and ideologies, are associated with the cultural politics of resistance politics (Levien, 2015, 2018). Development that disproportionally benefited the capitalists, industrial bourgeoisie, and other elites of several sectors has dispossessed the commons from their resource base (Bardhan, 1973; Baviskar, 1997b; Dwivedi, 2006; Alf Gunvald Nilsen & Roy, 2013). Development projects are administrative and bureaucratic practices of the State. It is how the State establishes hegemony and creates a 'world of meanings' (Mbembe, 1992, p. 2). Yet this meaning is to secure the models, practices, and discourses of State developmentalism, but development and its fallout – displacement and dispossession – are linked with the cultural politics of the State, and, therefore, it is a question of the 'political authority'. Dispossession can be mystified, as Levien (2015) mentions that coercion, material compensation, and legitimacy are the basic means to mystify the development cost. However, the whole issue of development is often rationalized in terms of 'poverty reduction' (Moose, 2007) to mystified dispossession and developmental costs. However, it is a clear fact that development-induced dispossession has turned many people impoverished (Cernea, 1988, 1990) and also marginalized the indigenous people, peasants, farmers, and industrial workers and caused a reduction in labour, social, and environmental conditions on a global basis termed as 'the race to the bottom' or 'global pillage'(Brecher & Costello, 1994).

The State accelerated the expansion of the commodity frontier across the world which restructured the production process. The production and extraction process is now organized at the commons' ecological costs (Lane & O'Connor, 1995; Alier, 1994). Unlike the other modes of production, capitalist development formed a commodity consciousness that multiplies with practical monetary value to the resource in the economy (Escobar, 2006b, 2011). It gradually co-opts and assimilates the alternative mode of production and polity. It also forms a new kind of social structure at the regional level in which capitalist relations are embedded and the way it produces a new kind of destitution and dispossession.

1.5 Analytical Pillars of Political Ecology

In this section, I talk about the political ecology's analytical pillars that help us understand and explore the events and link them with the theoretical mode of the theory. In this vein, four analytical pillars can be identified in the political ecology literature on resource access struggles that will also be employed in our research to explore and understand the conflict-related phenomenon.

First, political ecology treats the struggle over natural resources as the struggle to defend livelihoods. It has cultural, economic, and ecological dimensions related to different discourses and worldviews regarding the use of nature (Escobar, 2006a), unequal distribution of resources (Biersack & Greenberg, 2006; Li, 1999; Robbins, 2014), and environmental transformation (Li, 2010; Moore, 1993), respectively. The political and cultural significance articulates the values, interests, and actions. In this vein, it is essential to understand that livelihoods are not economically driven and structured but have cultural and political significance (Bebbington, 2013).

Second, political ecology observes conflict through the concept of accumulation through dispossession (Harvey, 2004). In this vein, some of the other concepts, that is, power asymmetry (Bryant & Bailey, 1997), talk about the actor's desire to control or access nature which induces conflict. Escobar's (2006b) concept of assimilation argues the practices of incorporating the local population into modern discourse, processes, and development practices. Here, assimilation is a practice connected with the politics of the development intervention and mainly deals with the cultural dimension of the conflict, resulting in what (Padel et al., 2013) is termed 'cultural genocide'.

Third, it generally considers the strategies of the resistance, which Scott (1989) refers to as an everyday form of resistance. Scott also terms the concept as 'the art of not being governed' (Dove, Jonsson, & Aung-Thwin, 2011), which is similar to the Gramsci concept of counter-hegemony (Mann, 2009). The art of not being governed talks about the history of deliberate and reactive Statelessness, whereas counter-hegemony means resistance to the dominant hegemony's agenda (Hall, Morley, & Chen, 1996). Fourthly, contextual structural force determines State-induced conflict. Local people's struggles over access to resources are often linked with the economic and political processes (Blaikie, 1985; Li, 1999, 2010; Peet & Watts, 2004). This theme helps to understand the actor's ability to derive gains and the politics over access to resource use.

To explore the social change in structure, the study follows the theory of social formation and uses the Marxian resource to use framework (Madhav Gadgil & Guha, 2013). The transformation of the Indonesian upland as studied by Li (2005b) also helps to explore the phenomenon. The theory of access (Ribot & Peluso, 1999) assists in understanding how powerful actors' ability to access or get benefits deprive the Adivasis of their traditional and customary base. It also helps to know how powerful actors, through structural mechanisms, manipulate the laws and customs to maximize profit from the natural resources.

1.6 Methodology and Epistemological Issues

This study is guided by the framework of *Epistemologies of the South* given by Santos (2012). The study also follows the model of the extended case study (Burawoy, 1998) and reflexive ethnographic science. The extended case study method is used for data collection. Only a brief discussion of Santos's *Epistemologies of the South* and reflexive ethnographic science will be carried out in this study. By epistemology of the South, Santos means

> the retrieval of new processes of production and valorization of valid knowledge..., in a systematic way, the oppression and discrimination caused by capitalism and colonialism.
>
> (Santos, 2012, p. 51)

For him, unlike a region, the south is a metaphor where humans suffer from capitalism and colonialism and resist overcoming the suffering situation. He mentions that the construction of epistemologies in the South must be built in four steps: sociology of absences, sociology of emergences, ecology of knowledge, and intercultural translation (Santos, 2012, p. 52). This study only takes insight from the sociology of absence and emergence.

The sociology of absences refers to the population and the community who are silent, absent, and invisible in the dominant discourse of world-views. The sociology of emergences acts both on possibilities (potentiality) and capacities (potency) and replaces the idea of determinism through the idea of care. Santos argues that in the sociology of emergences there is a need to explore the work as a symbolic magnification of the movement, initiatives, and organization; for example, a small social movement or a small collective action that emerges against the hegemony. This study refers to Ho community, whose resistance is invisible and absent in the dominant discourse of development. Methodology refers to the general approach to empirical research of a particular discipline. It refers to a research technique that is more appropriate in a research context. This study is guided by the political ecology literature in which special attention is given to various aspects of sociocultural conflicts connected with the issues of development intervention, resource extraction, and mode of resource use. I address these research questions through a qualitative case study and ethnography. Ethnography in the political ecology discipline differs in critical respects from traditional ethnography. Firstly, the focus of ethnography is not on the description of the life-ways of a particular culture but as its principal object of study environmental conflicts and their multiple social and natural interrelations. Secondly, it does not focus on a single social group or culture but must deal simultaneously with numerous groups. Thirdly, the geographic scope is rarely limited to the local affairs of social groups, but invariably incorporates various levels of the social scale. Finally, while traditional ethnographies usually dedicated a chapter to the natural habitat of the group under study, in political ecology research the biophysical environment becomes a crucial element in virtually all aspects of the conflicts under investigation. It, therefore, requires special attention to the environment, its natural dynamics, and the relationships it maintains with social groups. Ethnographic research does not always come with the same potential as an anthropologist's fieldwork, even when it is more than a set of open-ended interviews. Participant observation makes us consider people's lives and social relations in their totality, always be open to considering new possibilities in an open-ended study, and constantly forces us to revisit our theories and assumptions. It is one of the core aspects of any qualitative research. The aspects that form the basis of participant observation can be observed for a long duration (long-term engagement), revealing the social relations of a group of people (understanding a group of people and their social processes), holism (studying all aspects of social life, marking its fundamental democracy), and the dialectical relationship between intimacy and estrangement (befriending strangers) (Shah, 2017). As the manifesto of post-structuralist political ecology, *Liberation Ecologies'* opening essay grapples with the State by focusing on its social constitution to exclude examining the resultant power of the constructed institutional form. The State was still primarily an effect of other power relations, not an object of study. Affirming the possibility of a dialectical state theory, and one that did not forget the need to iteratively perform the State's existence and powers, needed a merger with the Gramscian understanding of States articulated by *State Theory* (Jessop, 2009) and the aforementioned encounter with the sociology or ethnography of the State that came with *Colonising Egypt* (Mitchell, 1991) and *The Anti-Politics Machine* (Ferguson, 1994).

Though the study has a multi-sited field, this political ethnographic study follows the ethnomethodology concept of reflexivity. It goes with Bryman (2004, p. 543), who defines reflexivity as reflexiveness among social researchers about the implications for the knowledge of the social world they generate of their methods, values, biases, and mere presence in the very situations they investigate.

Aunger (2004) talks about the concept of context and how it is essential in data collection. Context and events are necessary elements of the reflective method. Burawoy (1998) also talks about the context and positionality of the researcher for data collection and analysis. Reflexive science is different from positive science. It is guided by inter-subjectivity and contextually. Reflexive science gives importance to the context and position of the researcher; therefore, it thematizes our presence in the field we study (Burawoy, 2004). In the present study, reflexive science is used as a guiding principle of data collection because the field we study can be only explored through the researcher's presence as a participant in the events and phenomenon.

The nature of this study is primarily qualitative. It is a dynamic process that links together problems, theories, and methods; a dynamic process is involved, which ties together problems, theories, and methods (Bryman & Burgess, 1994). A qualitative method is an approach which can be typified as concerned with an 'intersubjective understanding of knowledge, [an] in-depth approach, [a] focus on positionality and power relations, [and] contextual and interpretive understandings' (Limb & Dwyer, 2001, p. 6). The qualitative research method is chosen because this study mainly concerns politics, power, culture, and the population. This theme under political ecology also needs to understand, firstly, from the assertion of Neo-Marxism, which talks about 'the constraining effect of social, economic and political structures', and, secondly, from humanist approaches that study the mediating influence on spatial patterns and processes of human agency (Cloke, Crang, & Goodwin, 2013; Limb & Dwyer, 2001). A qualitative research method is also an interactive model (Maxwell, 2012), making the research design more reflexive. The activities of collecting and analyzing data, developing and modifying theory, and elaborating or refocusing the research questions depend upon the researcher.

1.7 Locating Field Sites

The present study problematizes Western Singhbhum of Jharkhand in India as a study area. In recent years, it has become a conflict site for several actors. State developmental programmes and accumulative projects have politicized the local ecology that can be observed in the everyday life of the Ho community of this region. Politicization and transformation of ecology had been going on earlier too; but it speeded up after the State's adaptation of neoliberal accumulative policies. The historical importance of the region (Kolhan Estate 1837) – the presence of the *Manki Munda* system (traditional governance system), fifth schedule areas, and ethnicity and feeling of nativity (Sen, 2018) for the *Kolhan* region – is an ideal situation that demands a study through political ecology perspectives.

Additionally, the ecological disequilibrium due to the expansion of neoliberal extractive industries and government intervention induces conflict and aggravates resistance that politicizes the local environment. The presence of several actors makes an ideal situation for research from the lens of political ecology. West Singhbhum became a site for multiple actors with different values and interests, and, hence, this would provide an ideal situation to fulfil the research inquiry.

Political ecology literature has a common theme of the failure of a site-specific selection. A critical insight that has emerged is that although local issues are essential in a broader framework, factors external to these issues are government policies and sociopolitical framework. The study uses ethnographic methodology to glean insights from West Singhbhum. The study focuses on the Hos of West Singhbhum of Kolhan. Kolhan is one of the five divisions of the Jharkhand. The district consists of plain and hilly areas predominated by the Ho population (DasGupta, 2011; Sen, 2011). Villages have been regarded as suppositions of much more comprehensive processes and relationships. A tribal village differs from a common Indian village in character and composition. It is mainly situated in a comparatively plain-hilly region, and the most obvious way to grasp a tribal village is to consider its geographical setting, settlement pattern, and size on the one hand and the social setting (including ethnic composition) on the other (Vidyarthi & Rai, 1977). Two villages, Nakahasa and Gitilore, from the Tonto and Noamundi blocks have been chosen purposively as they are the sites of accumulation, dispossession, and resistance and are directly affected by the extractive projects. However, these villages are only research bases (Shah, 2010) to look towards extractive projects, schemes, and activities. Looking towards State is something irreducibly separate from the village site. The State is analytically present in this work, albeit as a distant set of institutional actors, exogenous to the study site (Robertson, 2015, p. 459). The limitation of site-specific efforts is that the local people are only minor players in a much larger theatre. The lives of people in villages are strongly influenced by the decisions of the central government and the conditions determined by it: construction of roads; availability of rural credit, subsidies, and tax incentives; inflation versus stability of the national currency; raising or lowering trade barriers; laws governing labour practices; receptivity to foreign capital; and many more (Terborgh, 1999, p. 169).

In this context, it should be argued that political ecologists must leave the village and focus on State action (Bryant & Bailey, 1997). Peet and Watts have emphasized that, from the beginning, political ecology was 'not inspired by the isolated rural communities studied by Rappaport but by peasant and agrarian societies in the throes of complex forms of capitalist transition' (Peet & Watts, 2004). The idea of the fractured environmental State (Walker, 1989) demands a critical understanding of a destructive State. He argues about the destructive environment of the State and that it should be recognized at the time of State politics of development. This is because the roles and functions of the State have become kind of oxymoronic, which can be observed in the continuing potential for conflict between the State's roles as developer and as protector and steward of the natural environment.

1.8 Research Design and Data Collection

Research design is a blueprint of the study. This study is based on an extensive review of secondary sources, participants' observations, and fieldwork. Fieldwork was carried out in the West Singhbhum district during two periods: the first fieldwork (September 2015–February 2016) was exploratory and descriptive, followed by the second fieldwork (August 2016 to October 2016), which resulted in 31 in-depth interviews and seven group meetings in different parts of West Singhbhum. Apart from these two periods, a regular short trip of two or three days to the field had been undertaken almost every month from November 2016 to March 2018.

This exploratory and descriptive fieldwork was aimed at the following:

1) Gathering information on the ways the State formation and building process had been going on.
2) Searching the possibilities and opportunities to interact with the people to explore the local cases of everyday resistance and negotiation.

In the theoretical framework of the study discussed in the previous section, this study outlines the various aspects of the political ecology of the sociocultural conflicts connected with the issues of development intervention, resource extraction, and mode of resource usage. The issues of resource use and control, power, inequality, access, and justice are articulated in the research objectives of the study. As discussed in the previous section, I address these objectives through a qualitative case study and ethnography followed by the

Table 1.1 Using Extended Case Study Method and Participant Observation

Reflexive Method	*Construction of the Case and Ecological Phenomenon*	*Conceptual Strategy and Tools of Political Ecology*
Intervention and equilibrium	Description and review	Genealogy of development, State and development discourses, indigeneity
State intervention and disequilibrium	State policy, intervention, and change in livelihoods and lifeworlds	Five theses of political ecology, State formation, and changes
Process that induces structure of inequality, social forces, multiple-actor presence	Social process of resource control, expansion of capitalist economy, resistance	Everyday resistance, justice, exclusion, political ecology, and study of resistance
Reconstruction, shaping the ecology, resistance, and value indifference	Governance, assimilation, co-optation State-building programme	Counter-hegemony, the art to not being governed, degrowth, accumulation, assimilation, dispossession, and moral economy vs capitalist economy

Source: Author

extended case method. Table 1.1 shows how the extended case method will be applied in this study.

The study engages with the literature of political ecology and epistemologically engages with the 'epistemologies of the South' (Santos, 2015). The study narrates the voice of the local Ho community whose livelihoods and ecology have been threatened by the process of State formation, accumulative projects, and State-building programmes. Getting access to fieldwork sites and people involved in resistance was difficult. Johar, Birsa, Omman Mahila Sangathan, TRTC, and some key informants introduced me to these field sites. The standpoint of this study covers the experiences, narrations, and stories of the local population about the State projects and schemes to provide more focus on the effects of State formation and building programmes on Ho Adivasis.

1.9 Chapterization

The book is organized into seven chapters. As elaborated in the above sections, the introductory chapter delineates the scope of the present study. It also contains a review of the literature to understand the issues about Adivasis, tribes, State, accumulation, power, and dispossession in a broader frame of political ecology. Apart from developing a theoretical perspective from political ecology literature, this chapter also deals with research questions, methodology, research design, and discussion of ethnography. The second chapter provides a clear description of the field sites. It aims to discuss the historical background of the region's nomenclature and definition. It seeks to give a brief account of the village profile and the availability of resources and government services. This chapter also deals with the sociopolitical organization of the village. The third chapter critically examines various debates on State and development through the lens of political ecology. It aims to show how State development is merely a neoliberal project and capitalist constructed discourse and indigently is a political slot against the State formation based on terrorization, conservation, and accumulation.

The fourth chapter provides a brief account of the historical, cultural, and political background of the processes of change by explaining the genealogy of State formation among the Ho Adivasis of West Singhbhum through the five theses of political ecology. Theses of political ecology are used to discuss how State formation is tied up with ecological degradation, territorialization projects, and dispossession. The fifth chapter discusses what resources are meant for Ho and why they have been struggling and resisting the mining companies and the developmental plans of the State to save their resources. How can their resistance be conceptualized as everyday resistance? It also intends to show how the State's facilitation of the act of extraction and its illegalities are symptomatic of the power asymmetry that has an impact on Ho territory. The sixth chapter deals with how shadowy democracy is practised in resource-conflict areas of West Singhbhum. The last chapter concludes the study by providing the future direction and scope for further research in the field of political ecology.

Notes

1 The principal chief conservator of forests has not recommended the proposal of mining in the reserve forest. In letter no. 1134, dated. 21.12.2016, which PCCF submitted as appropriate decision by the Government of India but the state government has recommended the proposal of mining in reserve forest. For more, see p. 91, http://forestsclearance.nic.in/writereaddata/FAC_Minutes/111512121217108352016FC.pdf.

2 For more, see http://forestsclearance.nic.in/writereaddata/AdditionalInformation/AddInfoSought/2014_1_1118123512101TCN_3_.pdf.

3 See http://timesofindia.indiatimes.com/city/ranchi/Tribals-oppose-fresh-bid-to-construct-Icha-dam/articleshow/16554761.cms.

4 For more, see http://www.indiaenvironmentportal.org.in/files/file/first%20report%20on%20illegal%20mining%20of%20jharkhand%20vol.%202.pdf.

5 In 1938, Adivasi Mahasabha formed in Chotanagpur (now Jharkhand) was formed to put forth the claim for special status for the Adivasis. This study mainly uses the term 'Adivasi' because this study is on Ho of Jharkhand. Ho and Munda communities identified themselves as Adivasis. Their claim is based on nativity (Damodaran, 2006). This term is based on self-identification like the term 'indigenous'. They prefer to call themselves Adivasis, not tribe (Hardiman, 1987).

'Tribe' is the most commonly used term to distinguish them from caste groups having evolutionary and Eurocentric origins but the term 'tribe' is constitutionally recognized as the Scheduled Tribe. At places, tribe is also used in this book to underline the administrative perception and recognition. 'Indigenous' as a term was first used by the International Labour Organization (convention no 107) in 1957. I use it as a slot against accumulation and capitalism.

6 This study is not looking on the pre-state formation. The time of discussion mainly started after the establishment of a formal modern colonial state. This study enumerates how formation of the modern state is based on unified centralized system.

7 By using progressive contextualization, we can observe the actions or interactions of individual living things and can proceed to put these into contexts that make the actions or interactions intelligible by showing their place within complexes of causes and effects.

Chapter 2

Ethnographic Profile of West Singhbhum

2.1 Background of the Region

The Chotanagpur region of Jharkhand is one of the world's richest mineral zones and boasts approximately 40 per cent of India's iron ore. Due to its large mineral reserves, extraction of the mineral resources is the primary industrial activity of this region. This region is highly enriched in mineral production; therefore, in the colonial period, this region was also termed a 'highly mineralized zone' (Sen, 2011, p. 73). However, in the name of 'national interest' tied up with the metaphor of public purposes, the State is imposing gigantic industrialization and developmental process involving the exploitation of natural resources (Areeparampil, 1996; Levien, 2011). With 25.7 percent of the total iron ore (hematite) reserves, Jharkhand ranks second among all the states of India. Iron ore and limestone mining are one of significant mining industries in the region. Mining of iron ore in the Singhbhum area started at 1907 (Areeparampil, 1996). Largely, iron ore mining is currently going on at Gua, Jamda, Noamundi, Chiria, Manoharpur, Kiriburu, and Meghahatuburu. Most of the ore is consumed by Jamshedpur, Bokaro, and Durgapur steel plants. Limestone is another crucial and old mining industry in this region. At present, limestone mining is operating at Chaibasa, Nakahasa, Khutpani, and Jhikpani blocks of this district. West Singhbhum stands out as an underdeveloped region within the Jharkhand State in spite of these mining industries. It is historically evident that the establishment of the mineral industry in this region has a conflict of interest among different actors, including Adivasis, mining companies, and the State over access to these resources. Agriculture and dependency on forest product are the primary sources of livelihood. In contrast, the State facilitates mining and industrialization that contradict local people's ecological base and have become a significant sector for revenue generation.

This chapter briefly explains the region's ethnic history, nomenclature, and social organization. West Singhbhum, popularly known as *Pashchimi* Singhbhum, is one of the 24 districts of Jharkhand. This district came into existence as a real administrative unit in 1990, when the old Singhbhum

DOI: 10.4324/9781032637990-2

District, ancient as the Kolhan region, was bifurcated.[1] West Singhbhum is one of the three districts that constitute the Kolhan administrative division of Jharkhand State, having a specific kind of alternate production system compared to the Gangetic plain regions of India. It is between 21″58′ and 23″56′ North latitude and 85″ and 86″ East longitude. The entire geographical area is approximately 5198 square kilometres. The district is bounded in the north by Khunti District, in the east by the Seraikela-Kharsawan District, and in the south by the Keonjhar, Mayurbhanj, and Sundergarh Districts of Orissa, and in the west by the Simdega District of Jharkhand and the Sundergarh District of Orissa. West Singhbhum is situated in the south-eastern part of Jharkhand, falling under the eastern plateau and the hilly region of Chotanagpur. The major rivers of the district are Roro, Sona, Koyal, Bansloi, Kharkai, Icha, Gumani, and Baitarani. The centre of this district consists of the upland plateau edged in by a hill range that can be observed within a few miles from Chaibasa.[2] The eastern, northern, and southern parts of this district also consist of higher hills occupied by the Archaean group of rocks that resides vast series of highly altered rocks comprised of minerals like limestone, quartz, sandstone, mica, etc. This region, surrounded by low isolated rocks, is dispersed and covered with forest and hilly areas. The succeeding sections will nutshell the history, culture, and social organization of the Hos of this region.

2.2 Historical Background of the Region's Nomenclature

As far as the derivation of the name Singhbhum is concerned, there are three distinct theories related to the nomenclature of the name Singhbhum (Bradley-Birt, 1874; DasGupta, 2011; Sen, 2011). The first theory says that due to the large number of lions in the region, it has been named as Singhbhum. The second theory argues that ruling chiefs were of another race. They were mainly Rajput (a Hindu warrior caste group). They claimed that their patronymic caste group settled in the region, and the name Singhbhum came from their inherited Singh dynasty, which had ruled the Porahat region of the Singhbhum. The region had acquired its name much before the Hos, who came from the southward of the Chotanagpur upland. This theory was a kind of particular commune's derivation as probable as another (third) theory, but the Ho community contemptuously rejected this theory. Bradley-Birt (1874) mentioned Hos' position, who collectively claimed that Singhbonga (the supreme deity who created the universe) had made this region. This land has been the designation of the Singhbonga (Sun + God), the object of their great adoration. These views lead us to the nature of demography and polity. The Adivasis living in this place showed pervasive dominance over the territory. The region also represented a political-cultural space for the Adivasis. Adivasis residing in these areas have had many totemic and sacred representations of god and goddesses from the hills, rocks, forest, river, and animals. Therefore, Bradley-Birt (1874) mentioned that the land itself had been given the appellation of the 'Singhbonga'.[3] The cognitive and imperial project of the British treated this region as culturally, ethnically, and politically different from other regions. The imperial knowledge production steadily recreated the Hos' ethnicity. Section 2.3 will shed some more light on it and seek to provide an additional description of the Hos' history and the notion of territory.

2.3 Hos and their Territory

This region represented a cultural mosaic of the ethnic communities: the Oraons, the Mundas, the Santhals, and the Hos; but the most phenomenological facts allied to Ho Adivasis were about the entry, expansion, and colonization of Singhbhum. Originating from the stock of the Munda tribe of the Chotanagpur Plateau, Hos are believed to have settled first in the northern parts of Singhbhum; gradually, they dislodged 'Saraks'[4] and Bhuiyans and colonized Singhbhum and shared a common descent, political memories, and cultural or physical traits. These all helped them to engage with the surrounding territory. Ho ethnicity had been understood through distinct imagination of place, language, and polity (Sen, 2011, 2018). The word 'Ho' stands for three realities: one, a person in this particular Adivasi group; second, the Ho language; and third, people who constitute the entire group. Its etymological meaning is derived from the Mundari or Kolarian root '*hor*', or '*horo*', which means a human person. More than 90 per cent of the Ho population lives in the *Paschimi* (west) Singhbhum District of Jharkhand State. The two neighbouring districts of Orrisa and two of West Bengal also have a few Ho settlements. According to Census 2011, the Hos constituted 10.5 per cent of the total Adivasi population (roughly 744142 people) in Jharkhand. The total Adivasi population of Jharkhand is 70,87,068. The Ho community is the most populous and principal demographic group inhabiting its region (Das Gupta, 2011). However, they are not natural settlers of the region but had come from other parts of the Chotanagpur region (Sen, 2011; Streumer, 2016). The gradual expansion and colonization by Ho into this region forced other settlers like Bhuiyas and other hinduized groups to move out. In other parts of Chotanagpur, Hos have found possessions of the Saraks community, who were also the most known primitive settler in Singhbhum (DasGupta, 2011). The widespread assumption about Ho culture was that this society was homogeneous and able to advance customs. Their economy depended on agriculture, forest products, a barter system, and frugality of wants.

The economy has self-sustaining and self-containing features (Sen, 2011). This community developed a specific mode of production system which had made Singhbhum (Kolhan) 'a secure asylum' for themselves much before the first British colonial contact with the region during 1817 (Dalton, 1872, p. 164). The epistemic intervention of the British administrators, ethnographers, and missionaries virtually demarcate the Hos through the undependable information and facts on pre-colonial Hos' past. To govern and incorporate the Adivasis of Chotanagpur into the British State system, the British had focused on the otherness between Ho and Munda and others in terms of manners, dialects, dress, and appearance. Streumer (2016) mentions how a political agent of British, Roughsedge, tactfully observed the Hos' distinct deportment and feature. It is doubtless to argue that it was a kind of bio-politics to control the population. Hos' ethnicity had different contents like dialects, different polity, and governance systems in comparison to the others. Section 2.4 briefly discusses the colonial history of this region. The colonial history of this region helps us to understand the earlier territorial settlement and the State formation among the Hos of this region.

2.4 British Colonial Invasion and Early Settlement of the Region

The Adivasi society was already feeling the unhappy effects of Hinduization and alienation of the tribal rajas and zamindars of the area when the British penetration began. Both impacts were, therefore, felt at once, and both introduced foreign notions and foreign people into the area, in an influx which led eventually to the economic ruin of the people. The tribal unrest of 1831–1832 was a crude form of protest against these changes and these external influences. It was a gesture of despair (Jha, 1964).

In the earlier section, it was discussed that Hos were not the original settlers. They had already developed their hegemony before the colonial period. They had occupied significant parts of the southern part of Singhbhum and became a threat to the local Singh rajas of Seraikela, Kharsawan, and Porahat. The invitation of Porahat raja to seek British help to defeat Hos shows how Ho Adivasis were good warriors and idealist rebels. The British subjugation of the Hos or Larka Kols was undoubtedly violent and terrorizing (Sahu, 1985; Singh, 1978).

Consequently, Adivasi resistance movements in various parts of Chotanagpur began with the Paharias uprising of 1756–1773. The Kol rebellion of 1830 forced the British to change their strategies and treat Hos and other Adivasi communities of this region as different from other governed communities. The whole Chotanagpur region was gradually endorsed with further treatment from the rest of the Bengal Presidency. From 1833, several legislations were introduced to protect the customs and social worlds of the Adivasis of this region. The history of tribal unrest had resulted in a new legislative enactment in 1833, but the eventual occurrences of the rebellion showed a hostile reaction against the colonial State formation. The British military intervention and constant administrative expansion finally subjugated the Hos in 1836. British Raj introduced a Kolhan Government Estate system comprising 26 Pirs[5] of 622 villages in 1837. Following the continuous rebellion, Thomas Wilkinson, a political agent to the governor-general of India, introduced an exclusive system of self-governance (DasGupta, 2011; Puthumattathil, 2014; Sen, 1999; Streumer, 2016). The tribal village council was given the right under this system to function independently.

The Wilkinsonian system worked under the superintendent of the district head, now known as the Kolhan superintendent.

However, it was one side of the story. The imperial project was significantly projected and worked as before with some apparent changes in the way of administrative intervention. However, during the post-subjugation period (from the establishment of KGE in 1837 to India's independence in 1947), almost all successive British officials had much appreciation for the Hos' way of life, cultural values, and manners due to their sympathetic policy and Hos' unwillingness to accept the Raj control. The region was incorporated into the colonial State system as a part of the South West Frontier Agency (SWFA) in 1837, but legally excluded from the civic and general rules and regulations of the British Raj. Ho Adivasis had been recognized as a tenant of the British Raj, and a binary form of taxation was introduced for the entire population without making any distinction between the 'original settler'[6] and the later settler. This new policy was based on imperial project of primitivism (Cederlöf & DasGupta, 2016). The rent-paying agrarian system was imposed on Ho Adivasis to make them less independent and aggressive. This was done through 'imperial ideology of rule' (Chandra, 2013a, p. 138). This ideology helped the British shape and legitimize the region's political order. The incorporation of customary village leadership within the canvass of colonial rule was based on the outward trappings of participatory government that helped the British State to co-option and retention of the communal villages into the geographical boundary of the State. The administrative protection and guidance were part of an imperial project that justified primitiveness, wilderness, and savages for subjugation. Colonial State had nurtured the local community through procedures and instructions and succeeded in transforming their unruly way of life. These governance tactics and the introduction of broad statehood into the region made Adivasis a governed subject of rule and control. The imperial civilizing mode of the British sedentarized the Hos by inducing them to bow out of the earlier livelihood pattern. This was introduced for the first time after forming the Kolhan Estate. The first two settlements were based on the plough (*hal*). Eight *Anna* per plough was fixed as rent, which was the first rent-seeking behaviour of the colonial State; then, afterwards in 1867, stick measurement had come and gradually rent was increased to two rupees. After these two settlements, one qualitative nature of land settlement was done that classified land into *bera* and *bad*, and the uplands as *gora* and *bakai*.[7] Land became a legal site for revenue generation and cultural identity. However, the paternalism and 'protectionism'[8] broached by the British to keep the *dikus* away were not all that successful, although as Singh (1982) has acknowledged that these were the first ever sympathetic and more humane approach to Adivasi issues in India. Moreover, besides their isolationist, paternal imperial protectionism, the British also advanced their colonial interests: they reserved forest and restricted Adivasis' access to it for commercial exploitation, forced sedentarization and agrarian expansion, introduced monetary economy, changed communal ownership of land to the individual for taxation, and acquired land for industrialization. These not only contradicted British paternalism but also necessitated the assistance of non-Adivasis and encouraged their influx further (Corbridge, 1993b; DasGupta, 2011; Singh, 1978; Singh, 2006).

Moreover, the only 'development' policy and work undertaken during British Raj was famine relief, in a defused and fragmentary fashion (DasGupta, 2011; Singh, 1978). The colonial policy was conservationist (Skaria, 2001) and was based on the idea of civilizing the savage. In this connection, it is necessary to mention that British and Hos relations were based on dominance and subordination. This dialectical situation is relevant even today in relation to the post-colonial State and Hos. The following section of this chapter will explore how the post-colonial State operates in a manner that aligns with its agendas.

2.5 Formation of Jharkhand and Ho Desum

Jharkhand for the Adivasis was formed when the Indian government changed the principle of linguism and accepted underdevelopment as the criteria for State creation. The State of Jharkhand was created by carving out the Chotanagpur Plateau and Santhal Parganas from the erstwhile Bihar through the Bihar State Reorganization Act, 2000.

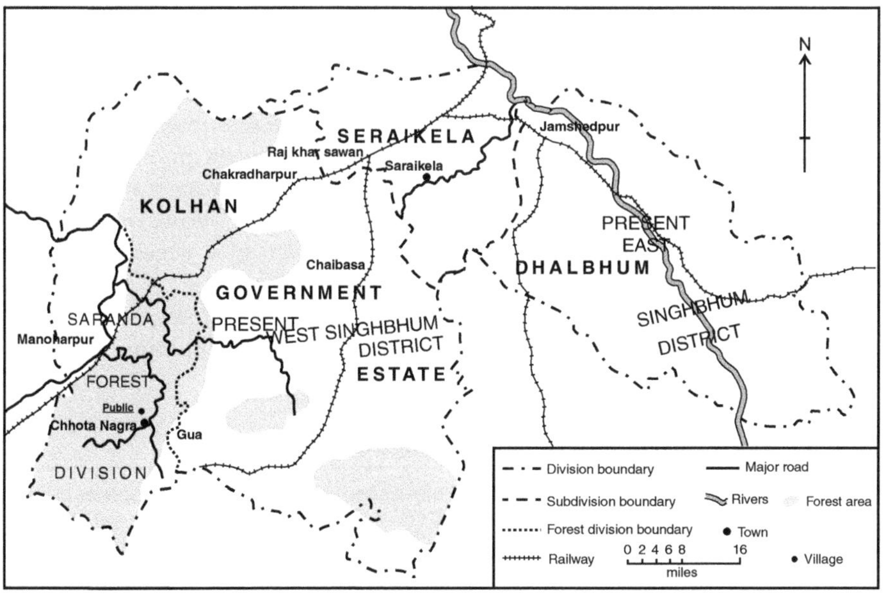

Map 2.1 Singhbhum District, its Forest and Plains in Kolhan Government Estate

Source: York, 1976, p. 19

Parganas from the erstwhile Bihar through the Bihar State Reorganization Act, 2000. Historically, it was argued that the tribal population of south Bihar (now Jharkhand) had been subjected to exploitation under the colonial and the post-colonial State. The formation of Jharkhand as a State can be associated with the ethno-national movement and the demand for self-governance. Restructuring of resource governance, institutions, and policies shows the complexity of State engagement with industrialists, which is evident in the State's pro-corporate policies. With the declaration of Jharkhand Industrial Policy in 2001, mining and industrialization were identified as crucial areas for the development of local communities in the newly formed tribal State (George, 2009). The State is home to extensive mineral resources, accounting for 37 per cent of India's total mineral wealth, including iron, copper, manganese, uranium, gold, and more. The Singhbhum-Mayurbhanj iron ore, known as the 'red corridor', runs through the district and boasts high-quality iron ore with over 60 per cent ferrous content (Karan, 1957). To generate economic surplus and create employment opportunities, Jharkhand State has expedited the process of granting mining leases and simplified the procedures for mineral production (George, 2009), (see Table 2.1).

Table 2.1 shows the region that appeared to be a financial liability to the British is still liable to the post-colonial State. In addition to the primary mineral, iron, minor minerals such as limestone, china clay, and quartz are also extracted and produced in the Singhbhum District. The mining industry's operationalization under the guise of development has made this area a focal point. The iron and steel industry initially acquired land for its mines in the district, with demands from companies such as Bengal Iron and Steel and Tata Iron and Steel being met under the Land Acquisition Act of 1894. The State was able to provide mining companies with virtually unrestricted access to village lands because industrial development served 'a public purpose'. Nearly 100,000 acres of land in southwest Singhbhum were transferred to the public domain for quarrying between 1915 and 1925. However, the local Adivasis could derive only a little benefit from such a general purpose (Corbridge, 1988; Kumar & Corbridge, 2002).

The priority of the newly made State to promote the mining and mineral industry come out very clearly in the region of West Singhbhum. After the

Table 2.1 The Production of Minerals During 2010–2011

S. No	*Minerals*		*Production in 2010–2011*	*Production in 2011–2012*
1	Major	Iron	2,32,25,000 MT	1,95,28,200 MT
2	Minor	Limestone	17,61,000 MT	17,60,000 MT
		China Clay	10,000 T	11,000 T

Source: Dept. of Mines & Geology (n.d.) http://dcmsme.gov.in/dips/Singhbhum%20West.pdf

formation of Jharkhand, the State government signed memorandums of understanding (MOUs) with transnational corporate houses to become a principal shareholder of the commodity chain. In 2005, it was reported that the Jharkhand government had signed 66 MOUs (Dungdung, 2015; George, 2009; Sundar, 2009a). The State rigorously kept up the 'civilizing' mission of the colonial time with a more discursive manner. The region has witnessed the longest tradition of the Jharkhand autonomy movement, which was based on the principle of self-rule and self-governance of the Adivasis. Though the movement was a politically successful one after the formation of Jharkhand, the cultural aspiration is still not fulfilled. The State that was made to fulfil indigenous aspirations has shifted its policies and chosen the path of capital-centric development. The State's establishment of industry, resource extraction, and generating royalties through minerals became tools and agenda. All successive governments of Jharkhand started to prioritize its mineral as a potential for growth, which shows how the State mediates between capital and corporate. It indicates that State is continuing to advance the colonial State's psyche in the name of development. The mega development projects and extractive industry (built for public purpose) by the State only focus on exploiting available natural resources for industrial development. Mining synonymous with development often ignored the ecological base of the local Adivasis. National development was not an answer to all the questions that a developmental project brings. The nature of the developmental intervention and its impacts on the Hos livelihoods and societal formation will be explored and discussed in Chapters 4 and 5 of this book.

2.6 The Contemporary Context

This section briefly describes the selected villages in which fieldwork was carried out. It mainly focuses on the demographic profile, resource base, livelihood patterns, and community societal and economic organization. It would help to set the panorama for the later discussion on resource governance, developmental politics, and the reason for everyday resistance.

2.7 Livelihood Trends: Changing Nature of Land and Forest

At the present, Hos are culturally embedded with the ecology for their livelihood. They are highly dependent on land and forest for their livelihoods. Farming and collecting forest products remain as main sources of livelihood in the Ho villages. Rice is the staple food among others, and it is widely used for brewing rice beer (*diyang*), and it is also used in rituals throughout seasonal and life cycles. People here are almost completely dependent on the autumn paddy crop. In the highland areas of West Singhbhum, autumn cereal and winter oilseed were harvested due to the hilly lands' light soil.

Table 2.2 West Singhbhum District Agriculture Profile 2010–2011

Total Geographical Area (ha)	528,870
Net Sown areas (ha)	58,992
Forest (ha)	109,740.08
Fallow land	61,029
Cropping intensity	Mono crop
Area high-yield variety seeds	No
Land not available for cultivation	233,250

Source: NABARD, 2014 http://nrcddp.org/resources_data/Dis100142.pdf

Mining impacts the natural drainage system, and gradually the region that was less prone to famine is converted into famine-prone areas. For Adivasis, the land is not only a source of livelihood but also life itself, both physical and spiritual (Mullick, 2004). Apart from food production, the land is valued for its ability to provide security, reinforce a sense of personhood, and offer access to credit, labour, and other exchange entitlements (Roa, 2009). The Hos are mainly a landowning community primarily engaged in agriculture with low technological inputs. Apart from agriculture, due to the high forest area, they sell firewood, collect forest produce, and sell rice beer (see Table 2.2). Land utilization is very average in these regions due to the lack of irrigation networks.

In the absence of assured irrigation facility, agriculture is primarily rain-fed in nature which only meets the consumption need of the local villagers. Table 2.3 reveals how West Singhbhum is very poor in irrigational facilities. This region is a government-neglected area, having very less agricultural development since India's independence (Hill, 2014).

Instead of promoting agriculture and traditional means of livelihood, the State is giving priority to the mining sector owing to the huge availability of natural resources. The State pushes economic capital into the arena of ecology to transform the ecosystem to generate immediate surplus and growth (Kumar & Mishra, 2017). The declaration of Jharkhand Industrial Policy 2001 allowed land transfers from Adivasis to mining industries in an accelerated manner suggesting the nature of development being facilitated by the State for the commons, how State interventions tend to change the ways people engage with the land and the forest and with each other. The result of this kind of symbiotic relationship between the two will be explored in the Chapter 3.

2.8 Village Profile (Research Base)

The initial survey of the village site in March 2015 and September 2015 led to the selection of the two villages for the study, situated in the low and the upland of the district. Selection of two villages is the only research base from which this study looks toward the State, which is already discussed in Chapter 1.

Table 2.3 West Singhbhum District Irrigation and Its Sources (Blockwise) 2016

Sl No.	*Name of Block*	*Source of Irrigation and Area Irrigated by Different Sources in the Blocks of West Singhbhum*															
		Canal		*Tank*		*RLI*		*DTW*		*STW*		*ODW*		*Others*		*Total*	
		No.	*Area*	*No.*	*Area*	*No.*	*Area*	*No.*	*Area*	*No.*	*Area*	*No.*	*Area*	*No.*	*Area*	*No.*	*Area*
(1)	*(2)*	*(3)*	*(4)*	*(5)*	*(6)*	*(7)*	*(8)*	*(9)*	*(10)*	*(11)*	*(12)*	*(13)*	*(14)*	*(15)*	*(16)*	*(17)*	*(18)*
1	Anandpur	0	0	0	0	0	0	0	0	0	0	0	0	0	0	0	0
2	Sonua	0	0	0	0	0	0	1	4	0	0	0	0	0	0	1	4
3	Bandgaon	167	154													167	154
4	Chakradharpur			388	155											388	155
5	Manoharpur	0	0	0	0	0	0	0	0	0	0	0	0	0	0	0	0
6	Goelkera	0	0	0	0	0	0	0	0	0	0	0	0	0	0	0	0
7	Gudri	0	0	0	0	0	0	0	0	0	0	0	0	0	0	0	0
8	Chaibasa	29	5	0	0	0	0	0	0	0	0	0	0	130	32	159	37
9	Khuntpani	0	0	0	0	0	0	0	0	0	0	0	0	38	13	38	13
10	Jhinkpani	0	0	8	0	0	0	0	0	0	0	0	0	0	0	8	0
11	Hatgamhariya	0	0	0	0	0	0	0	0	0	0	0	0	0	0	0	0
12	Tonto	0	0	0	0	0	0	0	0	0	0	0	0	0	0	0	0
13	Tantnagar	0	0	33	13	0	4	1	0	0	0	0	0	390	93	425	107
14	Manjhari	0	0	0	0	0	0	7	0	0	0	0	0	125	3	132	3
15	Noamundi	0	0	38	5	0	0	0	0	0	0	0	0	438	18	475	23
16	Jagannathpur	0	0	0	0	0	0	0	0	0	0	0	0	0	0	0	0
17	Kumardungi	0	0	18	0	0	0	0	0	0	0	0	0	49	10	67	10
18	Manjhgaon	0	0	122	9	0	0	0	0	0	0	0	0	0	0	122	9

Source: District Statistical Handbook, West Singhbhum 2016.

It helps to follow processes in and out of the field sites. Apart from these two villages, the study is also engaged with other connected regions to understand the reality of the social world of the Hos and West Singhbhum and also to understand State formation and building.

2.8.1 *Nakahasa*

Located approximately 18 kilometres from the main district headquarter, Nakahasa is a small village that comes under Bamebasa gram panchayat (see Table 2.4). This village/*hatu* lies in the forested interior of the Tonto block. Its name is derived from the soil used to clean the hair. This village is entirely an Adivasi settlement with 79 households. In fact, in the entire village, there is no other caste group. The villagers settled in according to *sasandiri* of the *killi*. Modern facilities and infrastructure are quite miserable in this village; there were only a few electric poles, but at the time of the fieldwork there was no electricity. Villagers mainly use solar lamps and kerosene oil for light at night. Power supply for domestic use and agriculture use was not available in this village. Hand pumps and tube wells were the only source of drinking water. During summer, many of the hand pumps were not functional. The sources of drinking water are a tube well and some operational hand pumps in the village. There is also a primary school in which only one teacher is permanent and others are para teachers. There are no banks, ATMs, or internet facilities in the village. ATM is approximately 10 kilometres far from the village. A road was being constructed at the time of the election. Things may probably change with the reductionist role of the State. A limestone mining company got a lease in this village but villagers protested, and at last the company moved away from the village. This village is nearer to the Ranjanka village, where the ACC cement company has been mining limestone since 1946; but to date, there is agitation against the management of ACC regarding employment and mining-induced agricultural loss. Resistance against the mining companies is too common in these regions. Hos of this village are traditionally engaged with their ecology, and they don't want any accumulative projects of State in their village territory. The nature of agriculture is subsistence, and villagers did not cultivate under market conditions. Agricultural technology is still traditional. They mutually help each other at the time of crop cultivation. They have a sense of belongingness and believe that land belongs to the village and that any accumulative projects degrade the life choices and harmony of the village.

2.8.2 *Gitilore*

In contrast to Nakahasa, Gitilore in the Noamundi basti block lies in the surrounding hilly areas of Saranda (see Table 2.4). Gitilore is a Ho-dominated *tola* (small village) settled after the Hembram *killi* (clan) displacement from

Table 2.4 Distinctive Features of the Two Studied Villages

Feature	*Nakahasa Village*	*Gitilore Village*
Distance from Chaibasa	18	73
Population features	Ho population	Ho population
Size of settlement	79	64
Total population (2011)	426	320
Administrative subdistrict	Chaibasa	Jaggnathpur
Administrative block	Tonto	Noamundi
Panchayat	Bambebasa	Noamundi Basti
Present administrative status	Revenue (khas)	Displaced from Korta village (no revenue village)
Marang *killi*	Bari	Hembram
Sex ratio	Female is more than male	Female is more than male
Literacy	50.5	48.2
Women literacy	Lower than male	Lower than male
Present administrative status	Revenue (khas)	Displaced from Korta village (no revenue village)
Ecology	Plain	Hilly
Livelihood	Mainly agriculture but also dependent on forest product	Agroforestry
Mineral availability	Limestone	Iron ore
Mineral zone status	Type B (backward)	Type A (major project with infrastructure development)
Infrastructure in village	Primary school, Anganwadi centre	Primary school, ASHA
Hospital	No	No
Electricity supply	Yes but many a time transformer failure is reported	No but in the Gundijora
Drinking water	Traditional mode (well, handpump)	River, nalla, and, recently, a water tank, which was installed by Tata
Road	Not concrete	Concrete
Communication	Poor	Poor
Land-holding pattern	Marginal to medium Dominant *killi* has a good quality of land	Marginal
Participation in contemporary struggle against developmental projects	Yes	Yes
Migration	Rare	Yes

Source: Author

the Korta village of Noamundi. The village of Gitilore is one of the bases of this study. It is the base from which this study explores the issues of culture, power, and development issues of this region. It has around 64 households with a population of about 320 people. Approximately, 70 kilometres from Chaibasa, it is situated in the hilly landscape of Noamundi of Saranda hilly areas. The settlement history of this village is different from that of the other villages. The Adivasis residing in this village were the early settlers of Kurta village. The case of Kurta villagers is very old. After getting displaced and alienated from their land by the Tata Company, the tenants of Kurta created and settled a tola in Gundijora village. They did it because their customs permit them to reclaim forest lands for the settlement. In Gundijora, the total area covered by forest is 208.6 hectares, and after clearing the forest land, the villagers of Kurta settled a tola. The village households have some paddy land. Gitilore village is not in the government census. This comes under the Gundijora village. The total population of Gundijora is 190, which included 64 households of Gitilore. The village comes under the Noamundi basti panchayat. The irrigation system of the village is good due to the availability of river streams and Nala. Like the Nakahasa, rice is also the chief paddy crop of this village. The original settler of Gundijora has more paddy land in comparison to the late settler of the Gitilore. The nature of agriculture was subsistence, and villagers did not cultivate under market conditions. Agricultural technology is still traditional. The surrounding forest is a clear livelihood advantage that can be visible in the day-to-day survival strategy of the Adivasis of this village. During the year, villagers collected sal leaves for making leaf plates, fruits and timber, which they sold at the weekly *haat* (market). Now, most of the surrounding region of this village and block is under lease for mining projects. Opportunities for wage employment have increased in the recent past due to the erosion of life choices and the traditional system.

There has also been a different nature of construction and manual work under the several government schemes. Earlier, this region was covered through the irrigational channel of the river and stream but due to the depletion of the river and the impact of mining, agricultural land is mainly unutilized.

Talking about the available facilities of the village, Gundijora village is electrified but in Gitilore there is no electricity. A road was being constructed under the MNREGA. Things may change, but MNREGA and government scheme job opportunities are available in this region. The infrastructure in this area is reasonable in comparison to Nakahasa. There is also a primary school that is not up to mark. Another availability of amenities in the nearby village is a private hospital run by Tata management for their employees, but with prior permission of Manki, primary treatment of local villagers is done here. It is only a treatment centre having facilities for emergency care. Apart from this option, health service is not easily accessible. There is no government hospital, but the Accredited Social Health Activist (ASHA) service is working there.

2.9 Sociopolitical Features of Hos Village

Unlike the caste villages, the Hos villages of Singhbhum are not fully compact nor entirely dispersed (DasGupta, 2011). Villages are mostly linked with the ecological space of the Hos, which regulated their cultural and spiritual life. The Hos villages in their totality are surrounded by a considerable wasteland, agricultural land, and forest land. The forest can be nearer or far from the village, but it is also an integral part of the village. Before the colonial State formation, it was said that any Ho cultivator could claim or reclaim any wasteland or forest tract but after the coming of colonial Raj, Ho settlers required the permission of the Mundas (customary village chief) to reclaim wasteland and jungle. The boundaries of villages are demarcated by the *sasandiri* (burial ground). Other bongas (sprits) also form the edge of the villages. Each *killi* is entitled to make its own burial ground but this is not for the service caste who resides in Hos village to provide service to them. A specific place is given to them but only after the permission of the local chief. The system of separate *sasan* for different *killi* (clan) has preserved the distinctiveness of the clans. Clans in terms of *killi* have had a sociocultural and economic role. It has an economic significance. In earlier times, the reclaiming of forest and wastelands for settlement shows the collective effort of an individual *killi*, which also helps them develop communal solidarity.

Additionally, it can be said that the social organization of the Hos is characteristically a village-based one and non-stratified in nature because at a certain socio-economic point, villagers (one or two *killi*) had developed a certain degree of group cohesion and solidarity but the village-based social organization of Hos was partially designed to ensure the control of the founders over the village resources (DasGupta, 2011, p. 22). Controlling and access of resources is a kind of egalitarianism that can be applied to the core group of the village settlers. A core group is not universal in this region (DasGupta, 2011; Sen, 2011). Unlike the Munda Adivasis, Hos were not the original settlers and reclaimers of the forest of the region. The northern part of this region of Kolhan was covered by dense forest; therefore, this region was colonized and reclaimed by the Hos as Khuntkattidar, whereas Khuntkattidar was neither popular nor universal in the southern part of this region. Unlike the Khuntkattidar, Marang *killi* (dominant family of the region with the same clan) and dupup parja (the original settlers) were so common. In the present study, Marang *killi* and Khuntkattidar will be used interchangeably to refer to the dominant and core settlers of the region. The core settlers mainly settled in areas covered with rocky hills and jungles. These features separated this region from the surrounding territories. Both are essential elements of the social organization of the village. Their traditional village council is headed by a Munda, who was the village's judicial and administrative head; this system is still going but within the State laps.

2.9.1 *Language and Communication*

In these two selected areas, all the villagers, irrespective of their *killi* and community, use Ho dialects for communication. While communicating with outsiders, they speak the corrupt version of Hindi. Students studying in schools have knowledge of bit Hindi and English. It was observed that due to the outsiders' impact and compulsion to access the State policies and programmes, they prefer Hindi language for communication. There is no internet facility in Nakahasa village. Young people are supposed to use the internet at the time when they are nearer to the district headquarters, whereas in Gitilore village people use the internet facility.

2.9.2 *Livestock*

Livestock is the main feature of the Adivasi economy. In the selected village, there are good numbers of livestock. In both villages, it has been observed that each household has a minimum of half a dozen hens and a cock. Some households of Nakahasa have cows but they do not use their milk. They keep cow milk for the calf. Adivasis use only cow dung for manure and for everyday purpose of the house. Ox is still used for plough purposes, and only the Marang *killi* or the core family settlers of the village have access to a tractor. The mode of agriculture is still subsistence in nature. Contrary to Nakahasa, villagers of Gitilore have no ancestral land for agriculture. They are more dependent on forest products and government schemes.

2.9.3 *Religion and Folklore*

The households in the study villages belong to the Sarna religion. Christianity's and Hinduism's impact can also be observed in their ritual practices. They have a strong belief in Singhbonga and village deities. They call outsiders *dikus*, and believe that they are dangerous for their culture. The composite of culture and ecology has strengthened their cultural activities. As this area is dominated by tribals, the people belonging to the other service categories are not allowed to be part of their ritual. Hos are known for celebrating the festival in the laps of nature.

2.10 Conclusion

This chapter examines the historical and contemporary social, political, and cultural context of West Singhbhum. The primary description of the selected study villages seems to indicate that developmental projects and State interventions impact the Ho community. The mining industry (major or minor) in the name of development has generated surplus and royalties to the State and other actors but it often impacts local communities. Lack of development interventions in the context of education, water supply, agriculture, and infrastructure in the region has supported the ideological claim of internal colonialism. Tables 2.3 and 2.4 show the empirical situation of the region. Lack of irrigation facilities and infrastructural backwardness is the feature of the region. State dealing with land in the form of a commodity is a technical and legal issue that enhances State formation. To start unravelling these issues and phenomena, the Chapter 3 will be a detailed description of the term 'development' and how it is used in political projects.

Notes

1 Three districts, namely East Singhbhum, West Singhbhum, and Seraikela-Kharsawan, have been carved out of the erstwhile Singhbhum District. In 2001, West Singhbhum was again divided into two parts. With eight blocks, Seraikela-Kharsawan District came into existence. At present, West Singhbhum remains with 16 blocks and three administrative subdivisions.

2 Chaibasa is the headquarter of the Kolhan region that comprises East and West Singhbhum along with Seraikela-Kharsawan.

3 A supreme deity of the Ho community.

4 Sarak were also known as Sarawaks, and they were copper smelters. They had considerable mining skill and were proficient metallurgists. This community was also the ancient miners of this region. For more discussion, see (Ball, 1880).

5 Pir here refers to a cluster of villages resembling the paragana.

6 Original settlers are known as Khuntkattidar. For more, see (Cederlöf & DasGupta, 2016; DasGupta, 2011; Sen, 2011).

7 *Bera* land was fertile agricultural land used for aghani (winter rice) paddy, and it was safe from drought as it was irrigated from the river streams, tanks, and springs, whereas bad land required a good rainfall. In the upland region, land was composed of light soils. *Gora* lands were roughly cultivated, whereas *bakai* or *bari* were cultivated; two crops, usually corn and mustard, were cultivated (DasGupta, 2011, pp. 207–208).

8 British protectionism, reflected in 'the Bengal Tenancy Act 1885, the Central Provinces Tenancy Act 1898 & 1920, the Chotanagpur Tenancy Act 1908, the Central Provinces Land Alienation Act 1918, covered only a few segment of Indian Tribals' (Singh, 1972, pp. 393–394).

Chapter 3

State in Development

A Critical Examination*

Thinking of development illustrates that the social life of an individual, community, society, or area is subject to multiplicities of interventions. The theology of development assumes that intervention is needed to reform the social life of an individual, community, or society. Development intervention has a social and ecological cost. Development is a series of events and actions as well as a particular discourse and ideological construct that demands examination. It functions as a hegemonic order to order, control, and represent the regions or populations in particular ways.

This chapter shows how development can be used for many different political purposes, including some, and perhaps most, the conflict with its essentially egalitarian ethic (a better life for all). It is necessary to understand and decodify the meaning of development when the cries of development echo all around in the forms of displacement, ethno-genocide, marginalization, and dispossession. This study is about the issues and questions regarding the development intervention mediated and facilitated by the State and the changes it has brought to the local ecology. The imbalance caused in local and regional environmental conditions due to the 'development' activities has been the bone of contention of the core areas of political ecology research.

Placing local and regional environmental problems due to development activities has been a core of environment–society research. The ecological or environmental change needs a 'chain of explanation' as mentioned by Blaikie and Brookfield (1987). They phrase:

> Starts with the land managers and their direct relations with the land (crop rotations, fuelwood use, stocking densities, capital investment and so on). The next link concerns their relations with each other, other land users, and groups in a wider society who affect them in any way, which in turn determines land management. The state and the world economy constitute

* An earlier version of this chapter was published in the article titled *State in Development: De-codify the Cultural Politics of Will to Develop the Adivasi of Jharkhand* in the open access journal *International Review of Social Research*, 9(2), 147–164.

DOI: 10.4324/9781032637990-3

> the last links in the chain. Clearly then, explanations will be highly conjectural, although relying on theoretical bases drawn from natural and social science.
>
> (Blaikie & Brookfield, 1987, p. 27)

Political ecology as a method and tool will rarely carry all these elements in a specific singular link. Research on this subject focuses on the way State authorities accumulate and extract natural resources through the economic pressure and regulation associated with developmental activities. Another dimension in the same research context focuses on the way local communities respond to this developmental activity and its institutional arrangement. The political ecology perspective of development focuses on the idea of 'trialectics' that helps us to understand the dynamic character of the state in the articulation of the local productive system of a particular society. The trialectics understanding rejects the binary and inflexible dialectical interpretation. State abilities to mediate and negotiate or regulate the socio-nature rejected the common dualism of state and society as well as of the Western dichotomy of nature and society. The strategic and relational understanding of the state is necessary because the state corresponds, reshapes, and evolves with nature and society and is also an integral element of socio-nature. Thirdering, or trilaectics, helps to understand the dynamic and politicized interaction between socio-nature and the state (Ioris, 2015). State's 'will to improve' society is based on developmental narratives or cultural scripts (Hoben, 1995) that demand a legitimate kind of intervention in the form of governmentality (Lemke, 2010). The three categories – the state, the society, and the rest of nature – form the trialectical continuum that is an integrally connected factor of some contested and relational reality (Ioris, 2015). The state is an integral element of the socio-nature that is existent and attached to society through its ability to mediate and regulate socio-natural processes. State co-evolves with socio-nature and society in a truly trialectical relationship in which processes of change (development) are simultaneously reproduced on nature, society, and state. This reproduction and contested form of change is the locus of political ecology. Political ecology research narrates the uneven reproduction of socio-natures through case study and participatory observation. Robbins (2012) mentions that political ecology describes how uneven power controls the flow of value from the environment and enhances the unwanted process of accumulation, within the shifting systems of political economy that perpetuate both. The political ecology of developmental research traces the contextual forces that constrain and direct outcomes, and write an explanation of those outcomes, in a way that deals with how value flows out of the landscape, through local communities to the other, and how the process of accumulation works.

Nature may exist outside society, but through state's social investment in natural resources, nature gets social life (Appadurai, 1988). The state does not live in isolation. State institutional power and force help in shaping the ecological condition of the region through territorialism. It came in the form of the social force of the state that is used in nature to make it more productive (Ludden, 2003). State's territorial claims and ability to invest in nature through social investment transform the natural resource environment. It also serves as a political and hegemonic tool through which a particular space, region, and community become subject to state description. Additionally, state's operational and political sensitivity to development depends upon its structural determination, its strategic orientation, and its relation to accumulation (Jessop, 2009, 2016). The intention of this chapter is to give more focus on the description of 'development' to understand the connected factors of the three categories that form trialectics. It gives a description of the concept of development engaged with state facilitation of capitalist development as a process and activity.

3.1 Development as Process and Activity

This study engaged development as an 'immanent process of capitalism' and an 'intentional activity' (Cowen & Shenton, 1996). 'Development' is a complex and multifaceted term that can be understood as both a process and an activity. As a process, development refers to the ongoing social, economic, and environmental changes that occur in a society over time. As an activity, development refers to deliberate efforts to promote economic growth, improve social welfare, and protect the environment.

However, the term 'development' is often contested and ambiguous (Peet & Watts, 2004), as it has been used to serve different purposes over time. The meaning of development has remained unchanged, but the concept has undergone several transformations that reflect changing ideas about the relative roles of the state, civil society, and the market in promoting development (Peet & Hartwick, 2009; Peet & Watts, 2004). Other scholars like Nandy and Visvanathan (1990) conceptualized development as an idea and a community which has served as a reason for the state.

> The idea of development has served many purposes in our times. It has served as reasons of state, as a legitimizer of regimes, as part of vision of good society, and, above all, as a shorthand expression for the needs of the poor. It has produced a new expertise and created a new development, new community of scholars, policy makers, development journalist, and reader of development news, development managers and activists – who together can be said to constitute the development community.
>
> (Nandy & Visvanathan, 1990, p. 145)

Ideologically, as a term, 'development' is constructed at a positivist level to signify qualitative change in all spheres of life. But ironically it demands an examination because the way it has been practised is highly contested and politicized. As a process, development has a root in the uneven evolutionary paths of historical capitalism. Historical capitalism transformed its feature into actually existing capitalism (Sanyal, 2007). However, development as an activity shows how state or its governing bodies and apparatus deliberately project specific models to enhance the historical process of capitalism. These activities are consciously dominant in the form of discourses, planning, and operation. Development was conventionally used to refer to qualitative processes of economic and social change that have been used by the state and other developmental agencies as a recipe for social change (Pieterse, 2000). However, the state sets arts, new developmental plans, policies, and agendas of prestige in the name of nation building (Nandy, 2002).

Development plans and policies are thus the sites of the contest between the place and state apparatus. Michel Foucault focuses on the field of knowledge, such as economics, and natural history, that is codified and classified and represented in particular periods (Gordon, 1991). The state represents these developmental interventions as objective and politically neutral; but Foucault shows that these areas of knowledge are socially, historically, and politically constructed. Social realities in these accounts are multiple and change according to the context. This illustration of development provides insight to understand how developmental projects territorialize resource-rich regions to control the resources and discourses. The wider systemic forces of the state have arranged and attracted private industrial players to enhance developmental projects that form a complex web of relations. The practices of development projects are based on domination (Escobar, 2011) that has politicized the operational process of development. Development projects should not be viewed in isolation as those are politically constructed and are perceived as an ideal (Baviskar, 1997b; Dwivedi, 2006).

The rhetoric of development as a part of the neoliberal project is linked with the wider nation-making developmental project in which the state's role is very important to understand the full correspondence between the developmental responses of the state and the socioecological interest of the community and other classes. Indeed, Mildavin (2007) argues that the entrance of coercive states and new markets results in the appropriation of communal capital from the locals and into the hands of the non-residents and distant parties. Moreover, placing the cost of the development at the feet of the local community, where the developmental project was once, came to deliver bunches of promise in terms of providing jobs, and eradicating illiteracy and hunger. Exploration of the ways developmental activity works and the ways it brings socio-economic changes became fundamental to the political ecology of developmental research.

Development was – and continues to be for the most part – a top-down, ethnocentric, and technocratic approach, which treated people and their culture as abstract concepts and statistical figures to be moved up and down in the charts of progress. As a linear theory of progress, it is bound up with capitalism and Western cultural hegemony (Adams, 2001, p. 6). It often works for economic growth. Development is interested not so much in the growth of an economy but rather in the conditions under which production occurs and the results that flow from it. In terms of conditions, development pays attention to the environments affected by economic activity and the labour relations and conditions of the actual producers of wealth—the peasants and workers who produce growth (Peet & Hartwick, 2009).

By taking insight from the above statement of (Peet & Hartwick, 2009), this study examines the conditions under which production occurs and impacts the environment by economic activity. Development as a process and activity has continued on a resource-intensive path. It has seriously disrupted the ecological stability of the local ecosystem. The issue of sustainability, environmentalism, and production of nature or social construction of nature is associated with the larger corpus of developmental issues (Castree, 1995, 2008). It argues in favour of creating capital-oriented markets for the exchange of natural resources and consumption. It requires privatization and commodification of resource use and control. Immediate resource use to generate a surplus to meet the global resource demand also transforms the traditional productive relation. It is in this context that understanding the development is necessary, because it ignores the pluralities of ecological relations (Padel, 2016; Parajuli, 1996). Its premises assume that once economic growth is achieved, then all sections of society will be equally benefited. This linear assumption ignores the heterogeneity and completely neglected the socio-historical and political factors of the region (Baviskar, 1997b). Additionally, it also induced ecological movements. The ecological communities which are at the receiving end of the larger development projects started to assert their voice against the ecological changes. In resource-rich regions of the global south, the seed of cries for development is linked with the environment and local communities, mainly the Adivasis and the peasants. Discussion of capitalism in the form of developmental projects and 'indigeneity' is important in this regard. Li argued that 'indigeneity'[1] is a political category and space that is co-merged with the discussion of capitalism (Li, 2000). As a label or class, they resist the expanding nature of capitalism and the market system facilitated by the state. As a political slot, they remained outside the market and capitalist system. Additionally, they can be positioned and understood as a counter-movement to capitalism. As a historical fact, it can be argued that tribe in the context of development is understood and treated as a political category with an understanding of primitive *vanvasi* and natives (Rycroft & Dasgupta, 2011). Development discourses reinvented 'indigeneity' which can be also observed as the 'invention of the primitive' (Kuper, 2005). In the global south, in general, and India, in particular, it is a recognizably distinct identity treated by different groups differently based on their embedded interest of articulation. The articulation of a tribe that resists the developmental projects provides a temporal imperative and a political aesthetic 'other'. Sometimes, they are literally called by others through a process of social self-recognition and 'interpellation'.[2]

As a series of events and interventions, development not only affected the environment but also reconstituted human–environment relationships. These developments raise a new set of questions about how environments are created, sustained, and transformed in such a context and the role of the state and corporates in these processes. Secondly, it also claims to be an investigation of the historical processes to examine the claims of changes in socio-economic process. These changes enhance ecological conflicts, resistance, and ecological movement that challenge the quick snapshots of development that is facilitated by the state.

In the social sciences, concerns about the environment in developmental discourse have been budding since the late 1960s, and the field of environmental history offers a powerful model for the political ecologists interested in studying change over time. The resulting changes in human–environmental interactions at local levels due to developmental intervention became a central focus for political ecology research. Political ecology research is interested in understanding the unintended consequence of modern development (Jewitt and Kumar 2000). The consequences and changes in ecology and human–environment relations need an empirical examination. It is true that varieties of approaches and traditions to the politics of development over environment and society are an eclectic body of research frame of political ecology. A significant change has occurred after the discovery of the environment in development that has been engaged with the discussion of sustainable development. Escobar in his seminal writing titled 'Constructing Nature: Elements for a post-structural Ecology' talked about the politics of sustainable development. He stated that the discourse of sustainable development and conservation of environment strategies plays a crucial role in a development projects (Escobar, 1996). The productive conditions are transformed by the capital. Sustainability, nature conservation, and environmental issues may be placed in the broader context of what Haraway (2013) calls 'the reinvention of nature'. Changing production systems, transformation, and changes that come due to developmental activity requires a description of development conversion into the practical frame of modernization.

3.2 Development as Modernization

Development projects as legitimized projects of the state, which began to be implemented after 1945, were numerous and varied. Capital investment is seen as the most important ingredient in economic growth to define development. The term 'development' has been the central concept of today's world. In modern times, for new nation-states, the core meaning of development was economic growth. As time went on, the elements of modernity, i.e., mechanization, technology, and industrialization became part of this concept. After the end of the Second World War, thinking of development was based on the notion of linear progression or a notion of a straightforward route leading upwards to the goals of modernity that can be seen in the work of Rostow's stage of Economic growth. Gradually, development broadened and encompassed by adding modernization, GDP, and economic growth. Development is also represented as modernity. A noted theorist, Rostow (1991) in his seminal writing 'Stages of Economic Growth', outlined a series of evolutionary levels to achieve the modern condition. Starting from the 'traditional form' as per Rostow, all societies had to pass through these stages of 'preconditions for take-off', 'take-off', and 'the drive to maturity' to achieve the final condition of 'high mass consumption'. These stages are characterized by high productivity, advanced technology, the service sector, and urbanization. The advocates of the modernization thesis proposed that development efforts are needed to tackle the problem of underdevelopment. Underdevelopment is seen as a historical condition and, through reorientation of its conservative norms and values, can move towards the capitalist form to achieve a developed condition. The state has invested hugely in capital-intensive infrastructure, i.e., irrigation projects, road building and heavy industry. Guided by modernization theory and focused on rapid economic growth, these projects have created profound social dislocations in the region with 'destabilizing' socio and cultural effects. Advocates of this school argue that the repository of 'tradition as a socio-cultural obstacle' in the road to achieving the goals of development. They conceived modernization as a cluster of interrelated social changes—industrialization, urbanization, commercialization of agriculture, and technological advancement—that produced a deepening differentiation of social structure (Lerner, 1958; Parsons, 1991; Smelser, 2013). The structural adjustment programmes were started to tame the imbalance of macroeconomics. This practice of centralized development policies gets criticized after the undelivered promise of high modernist development projects.

Failure in development projects has proliferated the critiques concerning development and its consequences. It has divided advocates of development into two groups. The first group believed that redefining development is necessary and that there is a need for 'alternative development'. They put emphasis on whether the state's role should change to address developmental failure. The state shifts its conservative role and facilitates inclusive neoliberalism in which inclusion, participatory democracy, empowerment, and equality have become more fashionable. Some scholars (Chambers, 2006; Harriss-White, 1997; Kothari, 1990; Sen, 1999) showed the failure of top-down development and mentioned that there is a need to redefine and understand the development through empowerment, participation, and inclusion. However, the second group argued that development as an idea did not work as promised (Sachs, 2010). Sachs thinks that 'the idea of development stands like a ruin in the intellectual landscape'. Delusions and disappointments, failures and dispossession, and marginalization and displacement are a common story and companion of the development process. As the rhetoric of development, empowerment, freedom, capabilities, inclusion, and participation co-opted with the capital-centric discourse of the market and due to that the potentiality of these narratives is always in question which demands an exploration of different representations of the development.

3.3 Different Representations of Development

The different representations of development carry the discussion of how development as a form of control, governmentality, and neoliberalism is embedded in the power of the state.

3.3.1 Development as Control

Through development, the state is able to impose its interest, values, and beliefs onto the other actors. It advocates that development process is a continuation of colonialism by other means (Kumar & Puthumattathil, 2018). As Apffel-Marglin and Marglin (1990) mention that the development process helps human beings to control their environment through the logo-centric mode of thought. This definition of development is embedded in the opposition between man and his culture on the one hand and the environment on the other (Apffel-Marglin & Marglin, 1990, p. 140). This school argued that capitalist development necessitated the transfer of surplus from periphery to core. It only vaguely specified the exact mechanisms of this transfer. However, the plunder of natural resources was part of Dependency theorists like Frank's analysis of underdevelopment (Foster-Carter, 1976). These theorists focused, almost exclusively, on 'unequal exchange' through trade or the repatriation of profits within multinational corporations (Amin, 1976; Baran, 1957; Emmanuel, 1972). They mainly focus on the lateral transfer of surplus from periphery to core, irrespective of how it was generated and how it had created dispossession and change.

Development as modernity and linear progression was dismissed as a historical and political phenomenon by the Neo-Marxists theorists. Neo-Marxists such as (Frank, 1977, 1978; Wallerstrein, 1974) adopted the approach of a political economy to show the negative effects of the development as a historical process. Their work challenges the evolutionary assumptions of modernization theory. They argued how in the name of development a forced dependency has been created across the globe. Their work comes under dependency theory that recognizes that one's development is based on another's underdevelopment. The colonial paraphrase of civilizing mission became modernizing mission, and it is used as a means to exercise power. Exercising power, in the name of development, is to control the regions and masses. The thinking of development as power to control required new understanding. In succeeding discussions, it is shown that development as discourse is a kind of governmentality. Through developmental projects, powerful actors imposed their interests, values, and beliefs onto other people (Crewe & Axelby, 2013). It is a continuation of colonialism. Development as control challenged the essential impression of modernization theory and argued that there is a need to understand the relationship between development and underdevelopment.

3.3.2 Development as Governmentality

The post-development theory came out after the failure of the developmental discourse. Discourse is a claim to truth. Developmental discourses show how a rule is accompanied. Robbins (2012) mentions that the logical extension to the notion of power/knowledge pervades a network of power relation that creates subject. It naturalized the market and process of capitalist development. People come to internalize the mandates of the state and the interests of capitalism within themselves. Robbins discusses the ways development activity, environmental management, and governance become normalized within communities. He says people come to self-govern, and in a way this insidiously subverts their interest to the interests of the state, to capitalists, or to powerful institutions (Robbins, 2012, p. 75). Other theorists like (Escobar, 2014; Esteva, 1992; Ferguson, 1994; Peet & Watts, 2016) worked to deconstruct the naturalization process of the development paradigm. They exposed the power politics and revealed how development as a discourse and practice serves the interests of the powerful actor. They advocated that development is not power-neutral.

Foucault's concept of knowledge, power, and authority discourse shows how development is historically and culturally shaped by power and domination relations. A close analysis of signs, structures, language, and institutional arrangements shows the ways developmental truth is produced at a cultural and a political level. However, post-development questions the very way the knowledge and the truth are produced within the developmental discourse. Escobar (2011) mentions three constituent elements of development discourse. The first group of elements are economic and material factors which include capital formation, technology, industrialization, and agricultural development. The second are cultural elements that carry education and modern cultural values. And finally, the third is the category of institutions, which include the World Bank, IMF, and national planning agencies. He mentions that development discourse is not only the combination of these elements but also the 'establishment of a set of relations among these elements' and systemization of them. It is conceived not as a cultural process (culture as a residual variable is to disappear with the advancement of modernization) but instead as a system of more or less universally applicable technical interventions intended to deliver some 'badly needed' goods to a 'target' population.

Development is to be seen 'not [as] a natural process of knowledge that gradually uncovered problems and dealt with them', but rather 'as a historical construct that provides a space in which poor countries are known, specified, and intervened upon' (Escobar, 2011, pp. 44–45). This developmental discourse constructs the specific truth, through which a region within the global south can be targeted, problematized, and managed in a certain manner that sustains hegemony. Escobar pursues a Foucauldian analysis of development discourse and defines it as follows: 'Development can be described as an apparatus that links forms of knowledge about the Third World with the deployment of forms of power and intervention, resulting in the mapping and production of Third World societies' (Escobar, 1992). These mapping and production are established through several institutions, socio-economic processes, forms of knowledge, technological factors, etc. Development advances normalities and abnormalities and labels them an anti-political machine (Ferguson, 1994). Illiteracy, traditional ways of life, subsistence economy and monetary poverty are labeled as the features of underdevelopment. Underdevelopment becomes a subject of political technologies, whereas development fosters a way of conceiving social life as a technical problem, as a matter of rational decision and management to be entrusted to developmental professionals (Escobar, 2011, p. 52). Escobar and Ferguson have adopted discourse analysis to interrogate developmental programmes, policies, and institutional structure of the state. They reject modernity and the self-claimed truth of developmental projects. The meaning of underdevelopment and alternative development is highly contested (Thomas, 2000).

> The very idea of development is under challenge to an extent not foreseen even a few years ago. Voices from the 'post-development' school claim that, at best, development has failed, or at worst it was always a 'hoax', designed to cover up violent damage being done to the so-called 'developing' world and its people.
>
> (Thomas, 2000, p. 3)

However, criticizing the post-developmental school, Pieterse (2000) mentions that post-development discourses are the radical reactions to the dilemmas of development. He mentions that the tools that they used to fight against modernity are also products of modernity. In this thread, Corbridge (1990, 1993a) also criticizes post-development for its equalization of modernity with universal science, bureaucracy, and even oppression. Additionally, he mentions that the use of modern science and technologies has substantial beneficial effects for society. But these critiques are illusive in some respects as post-development 'problematizes' modernity which is beyond the acceptance and rejection of modernity. It interrogates how power asymmetry is embedded in developmental projects.

3.3.3 Development in Form of Neoliberalism

The seminal writing of scholars Castree (2008), Jessop (1999), Kohli (1987) and many others argued that the rise and establishment of neoliberalism are associated with changes in the institutional architecture of the state. Defining neoliberalism is a tough task (Harvey, 2007) because it is not a mode of production and it connects a wide range of sociopolitical and economic phenomena at different levels of complexity (Filho & Johnston, 2005). The attempt to examine neoliberalism lies in the history of state capitalism. The roots of neoliberalism are varied and long. Further, its emergence is not dated as an event. It is difficult to analyze the manifest characteristics of neoliberalism at the time of its emergence. It fuses from a range of sources, including Adam Smith, neoclassical economics, monetarism, and the Austrian critique of Keynesianism. Its influence increased gradually with the breakdown of worldwide growth, the collapse of the Bretton Woods' system, and the degradation of Keynesian economics. The breakdown of communism and the cave-in of developmental alternatives in Third World countries helped it to spread all across the globe. Here, it is imperative to mention that neoliberalism and neoliberalization are not the same (Castree, 2008). Neoliberalism is a state of condition, whereas the latter is a process. As a process, neoliberalization shows how the state transformed its interests in favour of capitalists. The state did it through land accumulation and acquisition of land and resources. The dispossession of peasants, rural and indigenous people from the land, is considered to be a successful condition for neoliberal development.

Harvey, a noted Marxist of contemporary time explains 'neoliberalism' as a theory of political economic practices that propose that human well-being can be best advanced by liberating individual entrepreneurial freedom and skill within the institutional framework characterized by strong private property rights, free markets, and free trade (Harvey, 2005, p. 2). He additionally states that neoliberalism is a theory of political economic practices that proposes that human well-being can be best advanced by liberating individual entrepreneurial freedom and skill within the institutional framework characterized by strong private property rights, free markets, and free trade (Harvey, 2005, p. 2). At the discourse level, it has become more hegemonic. As a mode of discourse, it is incorporated into the common sense of the people that advocates that neoliberalism can be best to allocate resources and address inequalities. Buscher (2010) mentions that neoliberalism should be viewed as political, social, and ecological affairs that facilitate the dynamics of a capitalist market. Neoliberal structuring aims to expand capital accumulation and it is done through the financialization of capital. It is a kind of developmentalism where capitalist forces are stronger and more aggressive, and face minimal organized opposition than ever before. It only happens when the state facilitates the vision and objective of neoliberalism on common ground. Table 3.1 has summarized the above discussion and description of development. Section 3.4

Table 3.1 Changes in the Mode of Development Over Time

Period	*Approaches*	*Mode of Development*
1800–1900	Classic political economy	Progress, pre-stage of industrialization, evolutionism
1900–1940	Colonial developmental economic	Industrialization, resource management economic growth
1940–1960	Modernization Theory	Westernization, growth, social modernization
1960–1980	Dependency Theory (World System Theory)	Development of one at the cost of others; accumulation, authocentric, capacitation
1980–1990	Neoliberalism	Economic growth, structural reforms, deregulation, privatization, commercialization, social entrepreneurship, state as a facilitator
1990–2000	Post-neoliberalism	Crony capitalism, state as an actor and as an engineer

Source: Author

discusses how the state functions in a development that shapes the ecology and people relations within the existing ecosystem.

3.4 State in Development

Every state is known by the 'rights that it maintains' (Laski, 2010, p. 97). The right to develop the regions and communities became a primary task of the modern state. Focusing on the role of the state is important to understand what is intended by the development and how the state decides what type of development should be for the natural resource regions, community, and the ethnic people. In this thread, it is important to understand the role of the state who decides what development should be in resource-rich areas and what its results are going to be. Social investment in nature and its resources give nature a social life within the territory of the state which is guided by different developmental theses. Harvey (2014) finds it as a 'political economy of development as the mindless extension of capital's ecology into our life world' (p. 262). There is always a form of specialization and risk associated with this kind of development. This risk is conceptualized as interdependent development (Corbridge, 1982). Brookfield (1975) puts it as the real cost of the risk associated with the development. The risk associated with the development has been particularly strong in the Third World countries. States using development as a process and activity have served as a strategy of controlling the resource-rich regions by integrating them in a global capitalist system in which they are in a relatively disadvantaged position. Justifications have been often made for their disadvantageous position. Their position has been justified as necessary 'sacrifices for the nation's development'.

Sanyal (2007) deconstructs the development ideology and mentions how development as ideology is being in a systematic transition. For example, from its 1950s–1960s avatar of being 'development as a systemic transition' the discourse came to be known as 'development for improvement' during the 1970s (Sanyal, 2007, p. 88). Yet another view on development has been validated by scholars like (Bardhan, 2011; Chakravorty, 2013; Kumar & Mishra, 2017; Padel et al., 2013) and others who perceive the marginalization of certain sections as the 'necessary cost of development'.

Developmentalism in the global south is merely a neoliberal project and capitalist constructed discourse of the state. This discourse concerns not only economics but also promotes political and ideological conditions that enable capitalists to control other economic and political institutions in the name of nation integration and growth. Market fetishism is a central ideology associated with this form of state-mediated development, which only promotes the interests of capital. Development policy for indigenous peoples, from the colonial to the post-colonial period, has been a means of resource control of the region. This continues to this day through developmental discourses and projects that have been protested and resisted by indigenous communities. The state acts as a facilitator, protector, and advocate for the capitalist market economy, and the praxis of development becomes a form of governmentality (Lemm & Vatter, 2014). This type of governmentality defines development based on a projected desirable society, which is mainly concerned with capital-centric growth and linear progress models. The tenets of the linear progress model incorporate technical, economic, and ecological determinants. At the technological level, it focuses on efficiency and replaces traditional technology with scientific rationality. At the economic frontier, it refers to a change in the mode of production, generally involving the evolution from community farming to commercial or market-based production. On the industrial front, it makes a shift from organic power to mechanized power and enters into the phase of commodity-based exchange relations. Lastly, at the level of ecology, it reveals the transition from rural to urban. However, the state mainly focuses on all levels through the economic model of neoliberal development, which nurtures commodity fetishism, growth, and mechanization of labour. To facilitate this economic model, the state always restructures its apparatus and technologies of rule and produces a homogeneous developmental discourse. The discourse and the practice of development are based on the grammar of the market and the metaphor of national interests, which treat humans and non-humans as governed subjects. In the hegemonic discourse and practice of development, the Third World comes to stand for the category of the Southern 'local'; and the hegemonic can then define development in terms of a certain transition of the Third World, a transition bordering on its ultimate dissolution (Chakrabarti & Dhar, 2009).

In this context, development is viewed as a facade that serves the interests of capitalism. The state leverages technological advancement and knowledge production to facilitate the accumulation of wealth, which enables the capitalist class to prosper at the expense of marginalized communities. The resulting technological progress has transformed the forces and relations of production, and has created distinct categories for developmental issues. The state's interventions in underdeveloped and impoverished areas are guided by its understanding of these categories, and it is portrayed as being connected to nature through various constructed images. The state is simultaneously seen as both a representative of the interests of indigenous people and an advocate for capitalist growth. This paradox arises from two assumptions (Chakrabarti & Dhar, 2009). The first is capital centrism and the other is Orientalism. The field of Orientalism has historically depicted modern, industrialized sectors as superior to traditional sectors, and has advocated for ongoing developmental intervention to improve the latter. This approach has led to a dichotomy in the economy, with two distinct types of sectors. However, this view tends to prioritize Western notions of development, and perpetuates the idea that the West and North are advanced and civilized, while the South and East are undeveloped and undesirable. It is important to recognize that Orientalism is a form of political representation. For example:

- Western=urban=developed=rational=progressive=ordered;
- Non-Western=nonindustrial=rural=agriculture=irrational=savage/infantile

Capital centralism refers to the dominant and pervasive influence of capitalism, which is viewed as the central and inescapable reference point for discussing, regulating, and framing non-capitalist societies. The emphasis is placed on principles such as accumulation, efficiency, profit, and competition. This perspective is particularly evident in the colonial model of development, which was implemented in resource-rich regions of India. Section 3.5 talks about the colonial model of development in the resource-rich region of India to locate the politics and praxis of the developmental agenda of the current state.

3.5 Colonial Model of Development

British colonization brought formal systems to the tribal communities, creating a binary opposition between formal and informal ways of life. This incorporation of formal systems had two implications: people were either mobilized willingly or faced resistance and compromise, and a corresponding space was created. The British also imposed territorial identities administratively, which was not a tradition in the tribes. Before British colonization, the tribes did not have formal organizations, education was through observation and participation, land and forest were community-owned, barter was the mode of exchange, labour appropriation was reciprocal, conflicts were resolved through culturally approved ways, and personal relations were crucial in interactions. There was no written law or established procedures, structures, and rules. The concept of development was entirely foreign to the region prior to the arrival of the British Raj. It was only after the occurrence of several pervasive rebellions against the Singhbhum Raja's oppressive Zamindari system, characterized by an irregular increase in taxation and tribute (including *nazarana* in the form of cash and livestock), that the British made their appearance in 1830. Sensing an opportunity, the British extended their administrative system and framed these rebellions merely as a problem of governance. Over time, they also presented themselves as liberators of the Hos, leading to the establishment of the Kolhan Government Estate in 1837, which was placed under the direct administration of the Company's government. The establishment of Kolhan Government Estate had created new ties to this region with the outside and modern state (Streumer, 2016). As the Hos village community as a system started to integrate more within British Raj, their livelihood patterns and socio-economic organization started to deplete. The colonial policy towards the tribal regions was primarily focused on protectionism and isolation (DasGupta, 2011), but it also aimed to modernize the Hos and further the imperial agenda. The British pursued a contradictory policy of non-interference at the local level while enhancing the land revenue system and forest policy to generate more surplus for themselves. However, the introduction of a fixed and uniform rent without distinction between original settlers and later settlers, transformed the Hos from Adivasi to tenant, upsetting the socio-economic equilibrium of their society (DasGupta, 2011). A large-scale interference in the forest by the British Raj was to regulate, control, and tap the forest resources. British development activities were guided by the system of exploitation that consisted of two interlinked processes enacted by colonialism and outsider's (dikus)[3] encroachment into the territory and space of the Hos. The following folk song of Baiga tribe portrays the impacts of development that came in form of colonialism into tribals:[4]

> In this land of the English, how it is to live
> How hard it is to live.........
>
> (Cited in Elwin & Hivale, 1944, p. 316)

One of the important impacts of the colonial model was the modern state formation and the increasing control of the life worlds of the Adivasi people by colonialism, landlords/moneylenders, markets, and middlemen (DasGupta, 2011; Pati, 2011; Sen, 2011). They also mention that the problem of dispossession also started due to colonialism. It induced large-scale migration and forced labour. Singh (2002) mentions the Kol song that deals with the burden and pain of forced labour as follows:

> Alas! Under [the grind of] forced labour
> Blood trickles from my shoulders.......
> (Singh, 2002, p. 4)

Not only forced labour but migration also became a man-created displacement as a result of the colonial agenda of development. The migration of Adivasis to tea plant areas, mines, and factories was in a much-unpredicted manner (Pati, 2011; Sen, 2011). The nature of migration also stimulated the development of colonial capitalism. In fact, the question of integration and isolation was debatable under the aegis of colonial capitalism and state formation.

Migration, dispossession, tax increase and the state interference in day-to-day life of the tribe were also challenged by the tribe in the form of rebellions and movements. This movement was the tribal way to confront the reality of the colonial development model. Some of the uprisings were the Kol movement in 1855 of Chotanagpur, the Birsa Munda movement in 1890, and many others. The Chuar Rebellion was one of the earliest revolts by the tribal people against the British. The Chuar tribe, under the leadership of Bhagirath Manjhi, rose in rebellion against the British in 1767 and continued to fight for nearly 40 years. The British were finally able to suppress the revolt in 1805. The Revolt of Dhalbhum (Ghatsila) was another significant uprising by the tribal people in the early 19th century. The revolt was led by Durjan Sal and Raghunath Shah and was aimed at overthrowing the British and their collaborators, the zamindars and moneylenders. The Great Chuar revolt in 1799 was a continuation of the Chuar Rebellion and was led by Budhu Bhagat. The revolt was suppressed by the British after a few months. Tilka Manjhi, a tribal leader, led a revolt against the British in 1781–1784. He was able to mobilize a large number of people from various tribes and inflicted heavy losses on the British. However, he was ultimately captured and executed by the British. The Chero Revolution was a series of uprisings by the Chero tribe against the British in the late 18th and early 19th centuries. The Cheros were able to establish their own kingdom in the region for a brief period before it was crushed by the British. The Chotanagpur Tribal Revolt of 1807–1808 was a major uprising by the tribal people of the Chotanagpur region against the British. The revolt was led by the Oraon tribe and was aimed at ending British rule and the Zamindari system. The Munda Rebellion was a series of uprisings by the Munda tribe against the British in the 19th century. The first uprising took place in 1832 and was led by Birsa Munda. The second uprising began in 1867 and continued until 1890. The Kol uprising of 1832 was an uprising by the Kol tribe against the British. The Kol tribe, along with other tribal groups, were protesting against the imposition of new taxes and the forced labour system. The Santhal Rebellion of 1855–56 was one of the largest and most significant uprisings by the tribal people against the British. The rebellion was led by the Santhal tribe and was aimed at ending British rule and the Zamindari system. All of these uprisings played a crucial role in shaping the history and identity of Jharkhand. They were all expressions of the tribal people's resistance against the oppressive British colonial rule and their struggle for autonomy and self-determination. Today, the memory of these uprisings is still celebrated and remembered by the people of Jharkhand as symbols of their resilience and strength.

The movement and protest were against the increasing exploitation, erosion of customary institutions, and alien space formation. The Adivasis of this region drew the essence of a moral economy to challenge the model of colonial development interference. As discussed earlier, the continuous rebellion against colonial modernity and Raj forced British administrators to safeguard the Adivasi interest over land, forest, and local resources. The making of the Wilkinson law, the establishment of Kolhan Estate system, and Chotanagpur Tenancy Acts 1908 were such laws to safeguard the Adivasis' interests.

The post-colonial state also followed the colonial and capitalist path and laws (Sundar, 2009a). State undermined and crossed the protective boundary of Schedule 5 and Schedule 6 of the Indian Constitution and other legal provisions like Chota Nagpur Tenancy Act (CNTA) and Provision of Panchayati Raj Extension to Schduled Area (PESA)[5] to facilitate the developmental and state building programmes (Sharan, 2009; Sundar, 2009a; Upadhya, 2009). These acts and provisions are the products of popular struggle against the colonial development model, but now these laws and their manipulations are located in the developmental politics of the political economy of the state. The state takes advantage of internal contradictions within the law to favour a few industrialists and pauperises many Adivasis. Till the formation of Jharkhand as a post-colonial state within the nation-state of India, the following questions remained to be seen: Which 'public' do these acts address, and for what purpose have they been implemented? (Sundar, 2009b, p. 8). West Singhbhum of Jharkhand serves as an ideal site in many ways to understand state promoted capitalism which entered into an ecological phase where law manipulation, violations, and politics over access and control of resources, protests, and resistance against developmental projects and activities are going on that needs more elaboration which has been depicted the Chapter 4 of this book.

3.6 Post-Colonial Model of Development

In the post-colonial idea of the state, development popularly in the vernacular context is known as *Vikas*, which is accepted by the majority of the Indians as a legitimate activity promoted by the state (Rangan, 1997). According to Baviskar (1997b), the post-colonial model of development in India involves two interrelated processes: the immediate use of natural resources and the transformation of people and communities against their will into a hegemonized and dispossessed class. Kothari (1990) described four basic goals: firstly, the integration of diverse social structures into a homogenized discourse of national integration; secondly, economic development, which raises the standards of living for a larger section of society; thirdly, social equality in an egalitarian social order; lastly, political democracy in a culture that values authority based on status and power concentrated in the hands of the minority elite.

Additionally, in the case of developing resource-rich regions of Adivasis, the post-colonial Indian state's stated objectives of Adivasi 'development' policies and programmes focus on bringing them in national mainstream by reconfiguring them as margin, primitive, and backward and/or in need of development (Karlsson, 2004; Srivastava, 2008). The configuration and definition of Adivasis as a margin and their place in need of development is a part of the dynamic of modern state formation to restructure their relationship with the ecology and capital (Damodaran, 1997; DasGupta, 2009; Grove, 1990). Additionally, questions, like who rules or defines, whose interest is served, what constitutes the mainstream, why Adivasis or tribals should be brought into it, and what it means for them to be brought into the mainstream, have often been fundamental to the hearts of state system, institutions, and capitalist process for resource accumulation.

India's first prime minister Mr. Jawaharlal Nehru stated, 'No country can be politically and economically independent unless it is highly industrialized and the economy based on latest technical achievements of the day must necessarily be a dominant one' (Nehru, 1946). In his book titled *The Discovery of India*, he states the vision of modernized India in which the state would be the driving force behind economic production in the post-colonial era, but it can be said that Nehru's vision of development was also a reinforcement of colonial teleology of progress and modernity. India has witnessed the process of neoliberalism from late 1990. The political economy of development from the Nehruvian to the neoliberal market model has relocated the discourses and practices of development. After India gained independence from the British in 1947, the Indian state had two different plans for development: the Gandhian model of developing villages and the Nehruvian plan for prosperity through rapid industrialization. Nehru chose large-scale industrialization for the economic development of the nation, guided by the modernization thesis which is based on the positive effects of capital, science, and technology. The adaptation of the 'industrialize or perish' model of economic development is guided by the modernization thesis. The certainty of the modernization thesis was first critiqued by the dependency theorists because the crisis of modernization as a way to deal with development can be seen as a general crisis of modernity (Crewe & Axelby, 2013). This crisis of modernity has opened other possibilities for development but India has chosen the capital-centric path of development. The first Indian five-year plan focused on large irrigation and power projects, with little consideration for the social costs of these projects, such as the resettlement of displaced persons. The plan suggested the possibility of using a 'betterment levy' to finance the acquisition of new lands for resettling those affected by the projects. The second plan prioritized medium-sized projects and briefly mentioned social problems associated with large projects, but did not address the issue of displaced persons. The third plan acknowledged the uneven distribution of progress and reflected on the problems of implementing minor irrigation projects, but did not mention resettlement or environmental damages. The establishment of industrial estates and large industrial projects were seen as successful, with no mention of land acquisition or social and environmental concerns, except for pollution from chemical industries. Legislation was passed to enable the public sector to acquire new areas for coal mines, but soon it started to fall out in terms of dispossession, displacement, and other discontents of development. In India, by the 1950s, several legal instruments like land tenure reforms and 'territorialisation'[6] were being used more effectively to dispossess millions of people for dams, steel plants, mining, industrial townships, and public infrastructure. The capital-centric growth model to achieve the objectives of development was predominantly the developmental discourse of the post-colonial states but the intellectual roots of the modernization thesis started to fall after 1970.

As capitalism matured, the state's role as a provider was gradually replaced by the market discourse. Urban expansion and industrialization led to increased inequality and poverty. In order to modernize or develop regions, the state has resorted to various tactics to dispossess land and establish industries or extract resources. In India, like in other Third World countries, the primary task of development was to forge constructive relationships with capital (Amsden, 2000; Chibber, 2003; Evans, 1995; Kohli, 2004) and sometimes with labour (Chibber, 2003; Heller, 1996). However, the state's dispossession of land from peasants was never considered a significant aspect of this relationship, despite the fact that many high modernist projects (Scott, 1998a) entailed significant dispossession of rural land. Developmental projects impose the prerequisite conditions for capitalism to pursue the goals of Lakshmi (Rudolph & Rudolph, 1987), assuming that it will benefit all sections of society, but in reality it often leads to malignant growth (Bhaduri, 2016). The political economy of development in India, particularly in Adivasi regions of Jharkhand, has drawn attention because of the interaction between two different modes and means of production. However, with the rise of neoliberal resource accumulative regimes, the world's cultural diversity, consisting of numerous 'little' traditions, alternative productive systems, and organizations, has been absorbed into a linear assimilation. Additionally, as these processes continue, the idea of justice has been allowed to evaporate, posing serious challenges to the continued existence of humanity and the natural ecosystem. The presence of capitalist players and state politics of neoliberal development have transformed predominantly Adivasi territories into virtual conflict zones, increasing stratification and social differentiation (Shah, 2010). Deliberate misuse of the state apparatus goes on against those who resist and speak up against the neoliberal reign of development, violating their constitutional and fundamental rights (Dayal, 2010; Sundar, 2005, 2009a). The implementation of developmental projects and schemes is based on a web of domination embedded with the developmental politics of the state. The acquisition and lease of land and areas to extractive companies have displaced rural and agroforestry-based communities that have lived there for generations. These lands and areas serve as an ecological niche where local communities and other species fulfil their day-to-day requirements from the ecology. The state has a determining role to allocate and distribute the resources. Jharkhand is called a 'Ruhr'[7] of India. The resource extractive industry started in 1907 with the establishment of Tata Iron and Steel Company at Singhbhum (now in East Singhbhum)[8].

The demand for a separate state from Bihar is a kind of ethnic-regional movement that showed the failure of the post-colonial developmental model of India. This ethnic movement was the 150-year struggle by the Adivasis of South Bihar to restore their economic, political, and cultural hegemony over a region where they, the original clearers of the land, have progressively been displaced by non-tribal outsider and state for developmental projects. Industrialization and capital formation in the form of developmental intervention have developed a kind of internal colony in these regions where economic capital generates only few positive outcomes (Corbridge, 1988, 1993b). Section 3.7 talks about the continuation of neoliberal developmental politics and activity in Jharkhand that undermines the ideological base of bifurcation of Jharkhand from Bihar.

3.7 Post Jharkhand and Neoliberal Politics of Development

Jharkhand, a state that was separated from Bihar in 2000, had the option to pursue a development model that promotes indigenous ideology and living well with nature and society. However, instead of following this path, Jharkhand has chosen an export-led growth model and introduced mechanisms for social inclusion and politics of welfare to assimilate the Adivasi population. This approach has ignored the ideological base of the Jharkhand movement, which advocated for Adivasi self-aspiration and control over the region. The adaptation of neoliberalization in Jharkhand has led to deepening structural inequality among different sectors of the economy. The state has opened up valuable natural resources for immediate use, and new laws and corporate-friendly regulations have been enacted to favour corporates. However, this approach has had significant ecological impacts, including deforestation, displacement, migration, and land degradation. People have resisted the state's developmental ideology and projects, particularly mining industries, as a means to protect their ecological base of livelihoods (Sundar, 2005, 2009a). The Adivasi self-aspiration of 'Humara Raj', 'Swaraj', and 'Abub Raj' does not come into reality, and it has been left far behind in ideological space and discussion.

The resistance in these regions is a way for indigenous people to express their identity through issues such as water, forests, and land (*Jal, Jungle, and Jameen*)(Parajuli, 1996). This resistance also relates to the state's economic policies and its ability to control the market. The state has opted for this approach as it facilitates capital-driven growth and provides funding for welfare programs. However, this approach is based on the idea of linear progress and disregards the local traditions of communities. As a result, it reshapes society in a way that conforms to this model of development.

This new model of governance has been described as politics, not policies (Bebbington, 2000, 2010, 2011; Kaltwasser, 2011). The post-neoliberal politics of the state in Jharkhand goes beyond the Washington Consensus and intentionally enhances neoliberalism. Civil society and environmentalists have responded by adopting different mechanisms to address this trend. Despite Jharkhand's history of ethno-resource conflict and statehood movement, the post-neoliberal state aims to promote export of major and minor minerals while also introducing mechanisms for social welfare and inclusion.

Scholars like (Bryant, 2004; Bury, 2005; Cuveller, Vlassenroot, & Olin, 2014; Omeje, 2013; Shade, 2015) mention that just as in the post-neoliberal state, the resource extractive industry in the global south is becoming more aggressive and face minimal opposition. In Chapters 4 and 5, through empirical cases and secondary reports collected from field sites, this study argues how extraction activities and the rise of commodity frontier are linked with the state formation and building programmes. The study explores how the state's nexus with extractive industries enables them to employ ploys and tactics in the daily lives of the Ho Adivasis, a dominant community of West Singhbhum. This raises questions about whether development is for the community's well-being or capital-centric growth, where environmental and Adivasi issues are ignored. The extractive sites have become a key site of conflict, debate, and negotiation, and the state and other extractive players deal with these issues differently. Mining in West Singhbhum has divided people into those for and against mining, causing clashes and degrading the organic structure of Adivasi society. Both groups advocate for nature and the environment, but increasing resistance requires further exploration since there have been numerous disputes and resistance against developmental projects in West Singhbhum. The disputes between the state, mining companies, and local people are also related to the implementation of the development project, while ignoring local Adivasi issues of livelihoods and justice.

Defining post-neoliberalism as anti-neoliberalism (Sader, 2009) faces challenges in understanding politics and policies. In Jharkhand, the state claims to represent Adivasis but also invites foreign investment and appeases market players. The state recognizes customary institutions, celebrates Adivasi diwas, and sympathizes with Adivasi poverty, showing a complex character. State control over 'cultural signs' (Chomsky, Meyer, & Maldonado, 2010) is a post-neoliberal character of the state. As an ideology, neoliberalism is implemented and contested in policy and practice. Jharkhand is a particularly interesting point to study how neoliberal development as a process and activity deals with the local ecology. Jharkhand was established with the promise of a new ethnic development paradigm, but the state's development activities have been shaping the local environment and ecosystem, inducing change. The state's promotion of mining companies to extract natural resources has transformed the area into a conflict zone. The struggle over control and access to resources and rent distribution highlights the state's engagement, negotiation, and management of the ecology to promote commodity boom. Understanding state development projects requires analyzing resistance and protest against the state. As mentioned in the earlier discussion, looking at resistance and protest against the state is a vital element for understanding state development projects.

Notes

1 Indigeneity is also used here to represent *Tribe* and *Adivasi* that have an alternative imagination and way of life. Indeed, the epistemological position of these three are not the same but their imaginations, strategies, and approaches are the same and relevant against the colonization, capitalism, and state neo-developmental project. For more, see Rycroft and DasGupta (2011). In this study, it is used synonymously because as a class they resist the market forces over issues of livelihoods, justice, recognition, and ecology. It is already discussed in Chapter 1.
2 Interpellation is used here to show how *Adivasi* or *Tribe* is a subject of state power. The tribes are subjected to political webs of power. Their interaction with this web determines who they are even. Are they in favour of state developmental activity or are they the opponents? For more, see (Althusser, 1971).
3 Outsider exploiters include *zamindars* and moneylenders. Dikus refer to outsiders who are exploitative in nature.
4 Cited in (Elwin & Hivale, 1944, p. 316).
5 The Panchayati Raj Extension to Scheduled Areas Act, 1996 (PESA) gave wide-ranging power to tribes on matters relating to the socio-economic development of the community through the Gramsabha. Technically and politically, the act refers to extending the provisions of Part IX of the constitution to the schedule V areas, politically; it gives radical governance powers to the tribal communities and recognizes its traditional community rights over local natural resources. It not only accepts customary law governance and practices of rituals and religion but

also allows the traditional management practices of the community and gives direction to the state government to do arrangements regarding this.

6 State territorialization implies the making of a state claim over space to get or monopolize economic benefit. Resource control taxation access politics are the main features of territoriality. For more discussion, see (Vandergeest & Peluso, 1995).

7 Ruhr refers to resource-rich mineral region.

8 Before this industry, it was argued that mining in Jharkhand had started in 1774 (Carin 2013).

Chapter 4

State Formation Process and Change

An Exploration Through the Five Theses of Political Ecology

This chapter, in brief, throws light on the historico-political and sociocultural background of the course of changes in the predominantly Hos-dominated region of the West Singhbhum. It explores how development as a process and activity fueled through State formation has influenced and shaped the local ecology of the region. This chapter mainly starts with a brief description of the region that shows how development projects failed to deliver at the end, to the last person, as per the mandate of a social welfare state. Singhbhum is 'rich in minerals; Singhbhum is rich in the forest; Singhbhum is a labour-recruitment centre (Devi, 1981, p. 1597)'; but a government report shows that in West Singhbhum the spatial distribution of poverty is more than 80 per cent.[1] The top-down development policies and projects currently boosted by neoliberal policies and State's 'will to improve the region' through industrialization have failed to fulfil the aspirations of the local inhabitants. Today, mining in Jharkhand has become synonymous with development, and its effects in this district are remarkable. Politicians, State developmental officials, bureaucrats, and media refer to it as a means to achieve the development goal. Indeed, the 'State is leading producer of raw materials like copper, coal, iron, limestone, and mica' (George, 2009, p. 157), and West Singhbhum is one of the most mined districts of Jharkhand (Dungdung, 2015). Till 2011, 50 iron ore mining operations were on lease. As many as 22 approved areas for mining were operating only in the Saranda forest region.[2] Though Jharkhand is ranked second in the list of all the States with a reserve of 25.7 per cent of the iron ore in the whole country, the people of this State are still impoverished and deprived of basic amenities to live a dignified life. Jharkhand accounted for 20–25 per cent of the total steel produced in the country and is targeted to increase the production from 14.9 million tonnes in 2015–2016 to 25 million tonnes in 2015–2016. The future target is 4,596.621 million tonnes.[3] The haematite-concentrated region in Jharkhand has reserves of over 3,000 million tonnes. Out of these, 2,000 million tonnes of reserves are in the West Singbhum District.[4] As per the State's agenda of development, it requires more extractive developmental projects, which may

DOI: 10.4324/9781032637990-4

acquire more tribal land and forest. Despite this scenario, people are actively engaged in saving their land, culture, and ecology from the clutches of development projects.

Development through opencast mining, a feature of crony capitalism, has least prioritized the ecology. Extraction-generated capital is promoted as the only way to achieve developmental goals (George, 2009). To facilitate mining, State and capitalist players glorify the economic potentials of mining in terms of employment generation, infrastructure development, and the like. A proper social impact assessment has not been made due to the rent-seeking behaviour of the State and its inclination towards the mining companies. In the process, mining resources are antipathy to the people who, in the course, are marginalized and dispossessed. Their marginalization is a 'result of the State policy for promoting projects which is always an ongoing process' (Tsing, 1993, p. 90). Inclusive developmental politics and struggles over citizenship and the rule of law also enhance this process (Sundar, 2009a, 2015). The chapter shows how planned changes are intended to improve the conditions of the local people of West Singhbhum through the wilful political process of State formation that has brought social and environmental changes that either dismiss or undermine the local production system. Changes in the environment impact the overall system of society; in this context, environmental and social changes are complementary to each other (Norgaard, 2006). The process of ecological changes actively and historically produces social and physical milieus in the form of the transformation of ecology.

This chapter employs tools of political ecology to understand the phenomenon discussed above. Political ecology is characterized by the stubborn insistence on describing the transformation of the physical world. Core political ecology stories are the ones that describe 'environmental change and explain the conditions that affect that change or precipitated by it' (Robbins, 2012). Political ecologists follow these fundamental and linked assertions to understand ecological change. 'The costs and benefits associated with ecological changes that are not equally distributed among the actors which inevitably reinforce or reduce social and economic inequalities which hold political implications in terms of altered power of actors in relation to other actors' (Bryant & Bailey, 1997, pp. 28–29). It is necessary to place the issues of power and politics embedded in the State apparatus in a broader context of the environmental and societal changes that shape the discourse and practices of developmental projects.

This chapter deals with the State formation in Hos' territory through the lens of political ecology. Political ecologists have been engaged with various State formation theories while scanning for environmental conflicts. It ranges from neo-Weberian institutionalism to neo-Marxist 'regulationist' approaches to ethnographic works on the people and the area who remain inseparable constituents of State institutions. A chain of events unites them: the task of constituting the environment as a field for the State's legitimized important project, which seemingly may continue along with the increasing threats hovering over the access and control of a resource that induces conflicts. Nature is not simply the surface but an internal and integral part of the strategies on which the State acts. Political ecology's analytic tendency is used to understand the ecological changes of the region by focusing on the process of State formation.

4.1 Five Narrations of Ecological Changes

This section explains ecological struggles, livelihood alternatives, and issues of restricted access to resources through narrations collected from key informants during the field trips to the region.

Narration 1: Our relations with nature have historically been degraded. The intensive opencast mining resulted in deforestation, soil erosion, and dispossession. Continuous extractions of limestone, iron ore, and quartz have degraded water bodies through mining. Rivers and tanks are drying out (*nadi nale sab sukh gaye hai*).[5] Our dependency on forest products has been decreasing over time.[6] Drought and scanty rainfall have smacked our livelihood with the reduction in agricultural productivity. A tribal activist, Ramesh Jerai associated with the Johar, a human rights-based institution, mentioned that State-facilitated market penetration forcibly displaced their Adivasi system and turned it into profit-based capitalist system (*Rajya samarthit Baazar ka prabhaw badh raha hai jo humare vyawastha ko munafa adharit punjiwadi vyawastha ke taraf puri takat se dakhel raha hai*) (Ramesh Jerai; personal interview, January 2016).

Narration 2: Villagers of Gitilore protested against the extension of the lease area for mining by Tata under the banner of Bhumi Bachao Samanvay Sameeti'. Debram Tubid, a coordinator of the association, lament: 'What will we do if all the land and forest will be leased out to extractive companies in the name of development and public purpose'? Private companies and exploiters, locally known as *dikus*, have no concern for our ecology and environment. They only come for resource extraction, not for our development. They control our resources and push us to become a refugee in the land where our ancestors once settled. Natural resources are getting looted. Even State initiatives to promote industries in the dense forest of Singhbhum encroach on the wildlife habitat, including the elephant corridor.

Narration 3: Struggle over land and forest rights has been central to our history. The State has already leased half of its land to the mining companies. We don't want Mittal, Jindal, and Vedanta to come into our region. Once Birsa Bhagwan said, 'if we do not protest, nothing will remain'. His slogan against the exploitative kingdom was '*Abua raj seterjana, maharani raj tundujana* (Let the kingdom of queen be ended and our kingdom be established)'. It inspires us to fight against the exploitative colonial system. After independence, self-rule was a dream: *Abua Disum Abua Raj*' (Our Village, Our Rule). After a long history of ethnic and resource conflict, we got Jharkhand from Bihar in 2000. The State has always maintained latency. Our constitution guaranteed to safeguard the interest of Adivasis through the provision in the form of the fifth schedule, PESA Act, and CNT Act.

On the other hand, to our disclosure, the State often weakens these provisions in the name of national development and public interests. The government became an acquiring body for land, forests, and other natural resources like water, and the companies are the requiring bodies. Thus, it induced accumulation. The amendment in CNT Act[7] and leasing out of our land and forest to industries paved the way for destroying our culture and identity.[8]

Narration 4: We have no jobs; our diversified sources of livelihood are gradually shrinking. Out of distress, many of us have sold our holdings and got trapped in the exploitative labour market in constructive work offered in MGNREGA, Toilet Scheme, Pradhan Mantri Awas Yojna, and '*dhoba*'[9] construction. People are also migrating to district headquarters Chaibasa or sub-headquarters of the community development block, where diku contractors recruit them for temporary jobs. Seasonal migration is also in a rising trend among some young natives.[10]

Narration 5: This is not our Jharkhand; the alien administration co-opts our traditional institutions. Being Adivasis, we will not leave our land to the companies. Our region is resource rich, but outsider forces supported by the State made inroads to extract the nature-preserved resources.[11]

After the formation of Jharkhand, resource extraction policies and projects continued with more synergy. State developmental policies have been a hoax; in recent times, its hegemonic institutions and policies all facilitated companies to extract the resources, which led to unintended consequences. The mining history of this region is more than 80 years old. Now we understand the conspiracy of the companies. 'Through Gramsabha, we will decide the way we want to be developed'. '*Na Vidhansabha, Na Rajyasabha, Sabse bada Gram sabha*'.[12]

The region is the *Ruhr* of India, and people residing in this region are at the crossroads. They are fighting and protesting against the extractive projects. Their survival base is jeopardized. However, the more significant questions are still looming. The local people raise questions: Who has benefitted from the mining in terms of employment, education, and other amenities? Does it represent more significant developmental trends of State formation?

These questions are the hallmarks of political ecology. Mahasweta Devi, a well-known activist of tribal causes, has documented pain and suffering of local people through old metaphoric tales. She mentions that today the tribal existence in Singhbhum is symbolized in this simple song about a cowherd boy lost in the jungle (Devi, 1981).

> He can't go the eastern way;
> a tiger is there
> A big black bear waits on the south……
> (Devi, 1981, p. 1595)

She questions the situation of Singhbhum (Kolhan). She asks, what is the tribal going to do? Old mother Kolhan never answers. Her children would not beat the nagara when the outsiders enter Kolhan. How can she tell them what to do? What is done cannot be undone.

Nagara, a traditional drum, was used by the local villagers to get united in opposing and defeating the destructive alien forces. But as the State formation continues, they are co-opted with the everyday affairs into State operation and developmental agendas, which have also induced environmental changes. Changes here refer to a disturbance in the ecosystem and its relation. It has also caused inequality, marginalization, dispossession, and destitution of local people. For political ecologists, changes in a region can be understood in terms of changes in labour exchange relations or the mode of extraction by linking the process of accumulation and encroachment. It can further be interpreted as a 'moral economy versus capitalist economy' to unmask the reality of the particular region. The change in the form of ecological transformation after State formation in this region is driven by the internal changes and contradictions in the system and developmental projects. Contradictions are 'the essential elements of the system whose relationships to one another are necessary, but it inherently leads to crisis and changes' (Robbins, 2012, p. 55). Understanding the change required how society and State came across history and space and how State formation followed and impacted people.

4.2 Conceptualizing the Narratives: A Political Ecology Framework

The five dominant narratives reflect several issues related to environmental or ecological change with Adivasi marginalization in West Singhbhum of Jharkhand. The production and commodification of nature and related outcomes are seen in the shadow of significant explanations of ecological changes that the State once brought to justify development through the different extractive projects. These dominant 'narratives'[13] address various issues of environmental conditions; changes in the economy and productive systems;

responses to these changes and resulting impacts in terms of resistance, policy changes, or assimilation; and the emergence of new actors' interest due to interaction between two different sociopolitical conditions. The issues are multilayered and embedded with the sociopolitical conditions of the region. Linking political and ecological strands of the environmental changes may help to understand these narratives of socio-ecological changes. By focusing on the influence of the State (from the colonial to the post-colonial), capitalists, industrialists, and other actors in the transformation of local ecology, it can be argued that State's 'will to develop' (improve the region) is a kind of developmental politics that had an impact on local ecology. The trialectics include State, Society, and Nature (discussed in Chapter 3), in which the State's role is vital in reshaping and restructuring the socioeconomic condition of the region. State wilfully favoured particular classes and corporates that need a critical examination, especially from local people's narratives and perspectives (Robbins, 2014). Hence, the political ecology approach is apt to analyse the narratives of the Ho community as the type of a 'shared social view'. It takes multiple social and environmental conditions as a result of power at different levels. It tries to analyse critical discourses and policies relating to the value of 'traditional', 'historical', and 'local' ways of understanding and dealing with ecological processes.

In any sphere of society, socio-environmental processes take place through power relations. It can be material, discursive, economic, political, or cultural. These power geometries finally determine who will control and access the resources or other benefiting elements of the environment. This process shapes particular social and political configurations and also the environment. Besides, the concern is with the alternate polity and local people's resistance and adaptation techniques (Peet & Watts, 2004; Rosin, 1993). The political ecology approach can be helpful for understanding the State's politics behind the ecological change.

These dominant narratives describe changes, ecological struggles, and possible alternatives for the people to understand the livelihoods' alternatives dynamics in the changing scenario. Policymakers, State administrators, and other developmental agencies are primarily concerned with shaping the present and future of the region based on 'developmental narratives' or 'cultural scripts for action'(Hoben, 1995). Their modernist developmental efforts to improve the productive system of particular regions, places, or communities require an examination. After exploration of narratives collected from the field, this chapter conceptualizes these narratives through five dominant theses of political ecology to understand the changes that came in the 'form of development enclosure' (Robbins, 2012, p. 22). Table 4.1 deals with five theses of political ecology that conceptualize the region's environmental changes and ecological conditions.

The first narration, 'Degradation and marginalisation thesis', assumes that the productive system of the region or the particular society undergoes

Table 4.1 Five Theses of Political Ecology

Thesis	*What Is Explained*	*Relevance*
Degradation and marginalization	Environmental conditions (especially degradation and the reason for the changes through the explanation of 'Who' and 'How')	Ecological degradation, long blame on local people is shown in its more significant political and economic context.
Conservation and control	What are the outcomes, why it failed, political, economic, and cultural exclusion, 'Why and How'	Usually viewed as benign, having pernicious effects, sometimes it fails at its outcome.
Environmental conflict and exclusion	Access to the environment and conflicts over exclusion from it (incredibly natural resource)	Environmental conflicts are shown to be a part of more immense class struggles and vice versa
Environmental subjects and identity movement)	Identities of people and social groups, new or emerging ones (who, where, and how)	Political identities and social struggles are shown to be linked as fundamental issues of livelihood and environmental activity
Political objects and actors	Sociopolitical conditions (what impacts and How)	Political and economic systems are underpinned and affected by the non-human actors with which they are intertwined.

Source: Robbins, 2012, p. 22

transition due to over-exploitation of the natural resources. The exploitation of natural resources due to the integration of the local production system into regional and global commodity markets reproduces the local ecological system. The transition can also be due to the State development policies and acts in the form of interventions that facilitate environmental or ecological changes. These changes came under the transition of the innocuous local production system. It describes how contemporary development practices degrade sustainable community management and impose new alien institutions.

Shah (2013) has shown how the State formation among Adivasi regions has been instrumental in driving class differentiation, property accumulation, and stratification. She said:

> It was the State that brought in the outsiders in the early part of the nineteenth century, who then introduced a market economy and petty commodity production. It was the State that, in recent years, promoted education amongst the local population and created a few government-sector jobs reserved for Adivasis. It was the state that was creating wage work in the local economy. It was the State that was promoting petty contractor ships and the related black economy around development schemes. The State was

playing a driving role in sowing the seeds of class differentiation in these forests and hills in Jharkhand.

(p. 446)

Shah's study showed how the State introduced class differentiation, a market economy, and capitalist forces into the Adivasi areas, leading to a new kind of resource deprivation and poverty and pushing the natives to further marginalization. The main assertions of this narration have questioned the modernist developmental efforts and outcomes of the State intervention. She argues that developmental efforts have decreased local practice's sustainability and facilitated resource inequality among local resource users.

The second narration, 'Conservation and control thesis', postulates that the State's institutional efforts to preserve nature, community, and ecological sustainability, disables local community rights and control over the resources. Other studies done by Baginski and Blaikie (2013), Bryant (1992), Gadgil and Guha (2013), and Peluso and Vandergeest (2001) show that the State's efforts to preserve nature in the name of conservation and sustainability have uneven impacts on the local community and its institutions. State politics of conservation and development reflect hegemonic governmentality (Guizzo & Vigo de Lima, 2015; Sahoo & Kumar, 2017b). Related studies show that before the intervention (a kind of hegemonic governmentality), choosing a region as backward, static, and 'wilderness'[14] is a social construct (Li, 2005b). Control of resources and landscape has been seized from local producers in the name of the implementation of State laws on conservation and sustainability. It was done by labeling the local productive system as unsustainable, which once was historically productive and sustainable, as argued (Padel et al., 2013; Parajuli, 2010).

The third narrative, 'the environmental conflicts and exclusion thesis', talks about the politics of resource appropriation by the State authorities and private firms that induce politics of access. It argues that resource appropriation and enclosure accelerate or increase conflict between different stakeholders and actors. It also mentions that the environmental problem became socialized due to conflict between two groups. The term 'socialized' here means when local communities and groups secure control of collective resources at the expense of others. It also leads to long-term and short-term 'conflicts within the communities and between the State and the communities' (Robbins, 2012, p. 22). Mainly, resource development policies and conservation policies of the State are also used to 'ecologised'[15] the environmental conflict and issues.

The fourth narration of political ecology deals with the 'environmental subjects and identity issues'. Institutionalized and power-laden environment management, facilitated by the State, has formed a new kind of apparatus and systems, with their self-fixed and rigid definitions and understandings of the world and ecological ideologies and behaviours. This chapter aptly argues that people's beliefs, behaviours, and attitudes do not lead to a new environmental action or system; instead, 'new rule system or environment management lead to new kinds of people with new kind of understanding of ecology' (Robbins, 2012, p. 23). In addition to these new environmental apparatuses, conditions and regimes have also created opportunities for local people to represent themselves politically through political actions and behaviours (Ginger et al., 2012; Peluso, 2009).

The fifth and last narration of political ecology deals with 'political objects and actor theses'. It postulates that material characteristics of non-human nature and their components are associated with the human world. They also impinge upon the world of human struggles and are entangled with them. The rise of commodity frontiers and their connection with the global market leads to unwanted consequences of pernicious results. Resistance, unrest, and protest emerge from the traditional and progressive human and non-human alliances that are being marginalized, and due to such efforts, environmental subjects are also formed (Agrawal, 2005).

The above narrations of political ecology suggest several assumptions that help one to understand environmental changes and transformations. It provides a diversity of theoretical and methodological approaches to understanding the social system structured around the division of labour and power. It distributes differently the access and roles for the natural goods, and system is guided by a power asymmetry that ignores equal access to the common property. Power and knowledge embedded with the State intervention have interactive effects on the livelihoods and ecology of the local people. Political ecology assumes that socioecological relations share a common interest in questions related to issues of natural resource management, access, control, environmental knowledge, and their interactive effects on livelihoods and ecological change dynamics. Other assumptions of the political ecology suggest that property systems are a complex bundle of rights and are politically partial and historically contingent. It is also noted that historical experiences of the developmental project are not much beneficial for the local community. Development activities are rooted in the State's 'will to improve' the region in the form of governmentality (Li, 2007), which is poorly framed and has uneven results. It often results in changes that are linked with State formation. It showed how, as a hegemonic discourse, the State produces a new frontier while the pre-existing frontier becomes more marginalized.

4.3 State Formation, Power, and Production of Frontier

> To reflect upon history is also, inextricably, to reflect upon power.
>
> —Debord (1994, p. 134)

The process of State formation as a subject of historical research is not a new one. Sociologists, anthropologists, political scientists, and historians have referred to specific stages and theories that talked about the transition of political society from tribalism to a State system. The theory of State formation doesn't simply assume that State stands at the apex of society. As a central locus of power, the State manages all forms of institutional power and social relation (Gupta & Ferguson, 2002). It implies how the State acts as the social coordinator of the conduct and mechanism of the other institutions (Hansen & Stepputat, 2001). Noted sociologist Weber's concept of State advocated it as a political organization that claims and upholds a monopoly on the legitimate use of physical force in a given territory. Besides population, territorialization is a critical practical aspect of State formation. This study discusses how territorialization brought changes in the name of development that only legitimate and validate State power. The State is not 'the reality that stands behind the mask of political practice' (Abrams, 1988, p. 58). Accordingly, Sangampam (1986) mentions that State cannot be discussed without specifying a particular form of material production and its appended set of relations. Two questions always need elaboration at the time of the study of the State and its politics. First, what type of State exists, and where and in what kind of society does it subjugate? Second, how is the social product distributed among stakeholders and at what cost? State democratic form across the world is engaged with the capitalist ideology. State facilitation of 'capitalism'[16] has progressively absorbed the conventional little tradition of the community which has a different mode of existence. Here questions arise: How have democratic rules fueled by the capitalist ideology that facilitated outside organizations at the cost of customary rights been experienced since time immemorial? How did the State come to these regions, and for what purposes (Baviskar, 1997a; Puthumattathil, 2014; Sundar, 2005)? How did the State maintain its hegemony with dominance and without dominance (Guha, 1997; Nilsen & Roy, 2013; Sundar, 2005)? Dominance without hegemony stood that governable subjects came to rely more on force rather than consent.

This chapter mainly looks at State building history in the region from colonial times onwards. It deals with the changes in property rights and changes in land uses. It follows a brief description of the various development interventions regarding land acquisition policies, State control and institutional arrangement, and territorialization that facilitates new frontier types in the form of commodity and indigenous frontier. Frontier is not space itself; it is something that happens in and out of space (Rasmussen & Lund, 2018), and the political ecology of the State shapes it. The pivotal work of political ecologists has had an ambivalent relationship with the State and with what has been formalized as 'State theory'. The territory is critical and crucial to State formation and nation building. Sovereignty, laws, policies, and practices are means that represent the historical processes of uniformization of State practices. The control of territory is necessary for maintaining the State's economic governance. State governance in any sphere requires territorial planning and resource management that could be observed in any society where State formation occurred. It also provided an insight into how State policies shaped the given cultural, social, and productive practices. It is related to the trialectics of political ecology that have particular material and spiritual connections which shapes human and nature relation (Walsh, 2013, 2012).

In the State formation process, the State recognizes heterogeneous spaces and subjects related to territory, production, and economics. However, at the same time, geography or natural resource regions are in the process of being subordinated through the economic process that works to facilitate 'resource nationalism'.[17] State 'plurinational' approaches that recognize Adivasi socio-economic institutions and their customs as a means for their development are least prioritized in the pursuit of the neo-economic agenda. In Jharkhand, State politics of decentralization is a way that closely represents cultural and geographical differences. It is used to recognize State centric power structures and institutions that will be useful in the endorsement of the uniformity of the State (Becker, 2012). To maintain uniformity, State decentralization schemes and politics of democracy are operating at the village level. It also helps to address the resistance that came in reaction to development projects guided by capitalist accumulation.

The State often works to produce homogenized and standardized models of territory. Even though there is a lot of heterogeneity of issues, the State, through schemes and discourses, handles these issues to enhance the effective control of the State. These schemes and discourses are part of the State formation that has evolved within the nation's populations and geographies (Li, 2005a; Scott, 1998a). It also privileged overcoming injustices derived from economic power structures rather than injustices derived from their colonial power structures. Thus, Adivasi backwardness and wilderness of the region are framed as an issue of intervention of the State. The building of an organic State that integrates multiple issues and internal territories has the potential to stimulate extensive development based on a nationwide analysis of scarcities and needs. At the same time, it has deepened the problems of plural meanings and diverse understandings. The notion of State building avoids complex discussions of plural readings of territory and the use of resources and evades embracing diverse ways of life that constitute the milieu of socio-spatial relations in plural territories. Section 4.4 will give a detailed discussion of the region's history and politics around socioecological change.

4.4 History of Territorialization Projects of the Colonial State and Its Developmental Agendas

The policies of the colonial government postulated by the government officials taking social evolutionary theory as the base, for example to label certain social groups as 'childlike peoples' was unable to quite grow up (Skaria, 2001), started the projects of territorialization. Labelling certain groups as tribes or primitive or savages taking the concept of indigeneity as an inferior race running into the forest as soon as they come in contact with more advanced groups of agrarian people was on the basis of intervention and encroachment of the latter group, shrinking the territory of the former. This same approach was helpful to British officials in legitimizing control over the tribes and their territory. Indeed, the colonial State could govern the territory with a promise to develop the life of the tribal communities alongside a claim of protection from the outsiders (*dikus*). As discussed in Section 4.3, it can be said that territorialization is one of the crucial agendas needed for the project of State formation. This process of State formation in the tribal zone in India is comparatively an unexplored area of historical research. The historical exploration of State formation helps to understand the complex contemporary process and changes like tribal and non-tribal relations and interaction between little (inside) tradition and great (outside) tradition.

Singh (1971) mentions how State formation among tribes brought settled agriculture and the notion of generating a surplus. State formation among Adivasis of undivided Singhbhum requires a significant discussion on the position of the State. It shows that the State's coming in the form of stable polity and locus of governance in the Ho region was a reason for ecological changes and transformation. Panda mentions that ecological factors may be counted as a significant determinant in the development of the State (Panda, 2005). State formation and its journey from 'initial to mature' in the post-colonial era showed how State is essential to any discussion that deals with changes and transformation. The region was a frontier of the Ho Adivasi that gave them a sense of identity in the form of ethnicity. Colonial officials and ethnographers like Dalton and others documented their affinity for a territory popularly known as Kolhan and Hodesum (DasGupta, 2011; Sen, 2011).

The colonial State mainly treated tribes as the original population of India. Their policies and efforts were to treat and fix them politically in the form of populations' categories of savages, wilderness, and pastoral alike. Their efforts were to facilitate colonialism and civilize the wild. During the initial colonial phase, the prevalent view held by the colonial State was that tribes were "childlike" and unable to develop independently. It became an operationalized discourse that argues that region and population both require intervention in the form of development. It includes the process of commodification, territorialization, erasure of existing order, and the establishment of new patterns of governance and regulation. Territorialization and commodification are two different things and do not have a definite connection but both are used for territorialization (Ballvé, 2012).[18] In Colonial India, reserve forest lands and conservation of the same have not been a necessary example of commodification. It was a classic example of the transformation of a natural forest into a political forest.

In contrast, the land fixation, the initiative of mine opening, and the land acquisition policies may be cited as examples of the commodification of the resources. Both the examples of commodification and territorialization show how the State misinterprets the local customary and conservation rules and uses its power politics within its geographical territory. It also shows how State formation is a continuous process that works and is channelized through the power mechanism of different agencies of the State.

In the 'Preface' to his seminal writing 'A Contribution to the Critique of the Political Economy', Marx talked about the origin of the State. He mentioned:

> Legal relations as well as forms of State are to be grasped neither from themselves nor from the so-called general development of the human mind, but rather have their roots in the material conditions of life ... (which) constitutes the economic structure of society, the real foundation, on which arises a legal and political superstructure.
>
> (Marx, 1859, pp. 328–329)

Unlike Marx, Scott (2010) has broadly charted out four phases in the history of such traditional State-avoiding / fleeing people:

- A Stateless era;
- An era of small-scale states encircled by vast and easily-reachable stateless peripheries;
- A period in which such peripheries are shrunken and beleaguered by the expansion of State power; and
- An era in which virtually the entire globe is 'administered space' and the periphery is not much more than a folkloric remnant.

(p. 324)

Applying Scott's schemata to the history of Adivasi social formations, the second era might fall into the pre-Mughal period. The Mughal and Colonial periods might well be considered the third era. In this era Mughal and Colonial State and its power started to expand by conquering, defeating, attacking, and co-opting the local rulers and chiefdoms. Alternatively, it organized the local chiefdoms and their sacral polity, strengthening social formation among Adivasis (Fox, 1971; Thapar, 1999; Thapar & Siddiqieds, 1979). In the third era, the colonial authority established its power and extended its formal administrative structures, which entailed the gradual co-optation or subjugation of traditional autonomous Adivasi kingdoms and their territories. After the subjugation, forest reserves formed, transformed forest areas into settler land, and conservation started in the line of territorialization. Territorialization was one of the significant approaches of the State. Colonial and post-colonial States have adopted territorial approaches to controlling rural areas (Ginger et al., 2012; Peluso & Lund, 2011). But, in India, understanding these processes is quite different and complex. In Adivasi areas, the State used not only territorial approaches but also other tactics. State consolidation coverage with technocratic assertion automatically reproduces the traditional institution and organization (Sivaramakrishnan, 1999). It has been practised through the State's routine practices of demarcation, inventory, reservation, and regeneration.

The transformation of the 'natural forest' into the political forest and the fixation of land, initiative of mine opening, and land acquisition policies were representative examples of commodification and territorialization of the resources. Both the examples of commodification and territorialization show how the State uses its power and politics within its geographical territory. It also showed how State formation is a continuous process that works and is channelized through the power mechanism of different agencies of the State. This constant process shows how the State attempts to affect, influence, and control people and communities by delimiting and asserting control over the territory (Sack, 1983; Vollard, 2009). It is about including and excluding humans and their activities within the demarcated area. 'Territories are not frozen frameworks where social life occurs. Rather, they are made, given meanings, and destroyed in social and individual action' (Passi, 2003, p. 110). Territories are always manifestations of power relations and a product of State practices and discursive acts. The link between territory and power suggests that it is essential to distinguish between a place as territory and other types of places; it also shows how a particular State establishes a place as a territory through relations of rule and legitimate power.

The State often relies on its coercive power and politics to affect the geographical areas and people in a living space. It is a fact that very little-known material is available related to the pre-Modern period of Singhbhum. The region is diverse, fragmented, and surrounded by hilly and forest tracts. Talking about the history of Ho and their territory Hodisum (Ho territory) needs more research accounts (Sen, 2011). The engagement of scholars who have studied indigenous space, highlighting their attitude to natural objects, starts from colonial times.

O'Malley's writings gave a brief and sparse account of details about the region that mentioned that this region was politically and commercially significant. It was never in control of the Mughal Empire during the medieval period. During this period, the Saraks, Bhuiyans, Hos, Bhumijs, and others entered Singhbhum in search of habitat and area of influence (Sen, 2011; Streumer, 2016). Tickell was the first British ethnographer who gave the history of Singhbhum based on oral narration and survey in his memoir (1840); hence, the words Hos, Hodesum, and Ho language appeared for the first time. He mentioned that the ancestors of Ho came from the areas of the Chotanagpur Plateau that were inhabited by the Mundas tribe. In a recent study, Streumer (2016) mentioned that Hos, who spoke a dialect of the Munda language and shared many ethnic and religious beliefs and practices with Munda, once lived next to each other. It is observed that the names of the settled village and *killis* and clans of the Hos are the same as the border of the Munda settled village of Khuti and Singhbhum. With the passage of time, ethical indicators failed and the Adivasi groups started to consider themselves as a separate community (Streumer, 2016). Ho Adivasis, who moved southwards, found that the Bhuiyas and Jains colonized the region; however, Bhuiyas allowed them to establish and settle a village of their own, but after some time, there was a conflict on the issues of selection of the Rajas. Bhuiyas wanted their king, and Hos also wanted the same; in the end, they fought, and Hos subjugated and occupied the tract. This tract remained untouched to some extent. The geographical isolation of the Hos and 'their innate exclusiveness have preserved their tribal organization in its pristine form' (Singh, 1978, p. 10). The pristine form of incipient State formation among the Binjhals of Odisha can be a parallel example of orthogenetic development (Panda, 2005). The agrarian organization of the Hos follows in the wake of their tribal organization. The Adivasi characteristic of the Hos and the nature of the country they lived in moulded their agrarian organization to a greater extent. When the Hos migrated to Singhbhum as wanderers, they settled in the Kolhan, perceiving an abundance of fertile and virgin soil covered with the dense jungle and traversed by the various rivers and streams. They set themselves the task of clearing the jungle for residence and cultivation and thus became '*Khunt Kattidar*'.[19] Land, water, and forest were the loci at the time of village settlement. The basis of the Ho politics was the village panchayat. A village came into existence after the clearance of the jungle and demarcation of boundaries in the form of Jahira and stone burial. They were culturally very organic. It is because of the strong ties of mutual affection, born out of common hearth and home, common worship, and common sharing of the property under the fostering care of the same parents, knitted and together as one family. The injury to any member of the *killi* began to be viewed as injury to the whole *killi* and as such, was avenged by the whole clan (DasGupta, 2011; Sahu, 1985; Singh, 1978). It was very tough to sketch a separate line to distinguish their culture from politics. Both are embedded in their socio-magical form of religion that can be conceptualized as a sacral polity (Shah, 2010). The Ho village or *hatu* looks like a political as well as a social unit to a stranger, but, primarily, it was a social unit containing the members of a distinct *killi*. All important social, religious, or economic works were initiated and guided by the *killi* elders (Sinha, 1978). The communal organization of the Adivasis of Chotanagpur is singularly complete, the village being the political unit. Several villages, generally from 10 to 25, constitute a *Parha* or rural commune, under a Manki Munda system. They exercise executive and judiciary authority over the people in that region. They ran their distinct polity. They lacked a stable governance system, which was one of the reasons for their incorporation of the British system. Still, on the other hand, their institution and political organization also revealed indigenous growth. Every single village has a distinct polity. Their village panchayat was an 'ancient tribal institution with a strong democratic leaning' (Sinha, 1978, p. 20), having a feature of a little republic (Singh, 1978; Sinha, 1978). It would be out of place to mention that the political and administrative genius of the tribe found full expression in the democratic working of the village panchayat and the efficient discharge of the administrative duties of the Manki and Mundas: the two most outstanding tribal functionaries. The sketch of their organization was also talked about by colonial officials and ethnographers. One of the noted British officials, Dalton, who was also an ethnographer, talks about the pre-State system among the Ho Adivasis. He says that their political system was based on the hereditary headman of a united village system known as *Pir*. The pre-State system was not uniform among the Ho. Forest, hill, and inaccessible regions showed that Ho village setting was so natural, and it served as an obstacle to the uniform system of governance (Dalton, 1872).

Another colonial ethnographer Bradley Birt mentioned:

> roads there were none, only jungle paths from village to village. The forest was so thick that it was almost impossible for one man to walk along, and a way through had to be cut as they advanced. Range after range of hills had to be crossed where was slow and difficult progress.
>
> (Bradley-Birt, 1874)

Their village economy was based on a barter system with a less commercial link with the outside (Sen, 2011). It showed that their contact with the outsider was much less before the Colonial Raj. British, through their cognitive project of 'imperialism',[20] territorialized the Kolhan region. They identified Ho Adivasi as culturally, ethnically, and politically other. As the Western modernity was injected into the Ho territory, a slow and steady recreation of the region started. The areas where the British Raj secured legitimate control over territory was achieved in 1837 with the formation of the Kolhan Estate. After the formation of the Kolhan Estate, the British Raj started to uphold the monopoly on the legitimate use of physical force in this region. It was the time when the process of State building had started. British treated this region as a 'zone of the anomaly'.

For the people of Singhbhum, land is not just a source of livelihood; it gives a sense of connection, association, and 'belongingness with their history'. Their societal formation is subsistence and has a 'lineage-based mode of production'.[21] Their economy geared towards the domestic mode of production. The regulated land revenue survey and settlement system were introduced for the first time with the formation of Kolhan Estate in 1937 when the land began to be treated as a commodity that could generate revenue. The region was excluded from the general regulation of the British. Only after Wilkinson's report was submitted to British Raj did this region come under the direct control of Kolhan Superintendence. The region was regulated through the customs and people of the village. There was no civil procedure code, but it was a kind of British protectionist policy to address the uprising of the Hos against the British State building. The praxis in British protectionist policy for the Hos region was observed with the implementation of the imperial scheme of territorialization project and the introduction of the rent policy. It also helped the British government achieve their cognitive project through which they identified Ho as other than the civilized (Sen, 2011). It is a fact that the notion of the tribe gained a concrete function in India at the time of colonial State formation. Devalle (1992) strongly argued that 'there were no "tribes" in India until the European perception of Indian reality constructed them, and colonial authorities gave them their administrative sanction' (p. 51).

As time passed, British officials and administrators started increasing the rent. Rent was increased to one rupee in 1854 as it was eight *annas* per *hal* in 1837; rent was raised again in 1867 and was fixed to two rupees. The successive land settlements classified the land of this region based on its fertility. The qualitative classification of land into Bera, Bad, and Gora showed how the colonial State progressively engaged in revenue collection. In 1850, the colonial government learned about the availability of copper, iron, limestone, and manganese after a survey by British companies (Sen, 2011, p. 19). State's move to the extraction of iron ore, limestone, and other natural resources had produced a commodity frontier. A bunch of promises – to give employment and mining jobs to people – was first operationalized as discourses. Unbroken people's history and broken promises of the State have been part of the games bureaucrats played. The introduction of a capitalist economy through commodity exchange and market exchange was innovative for this region. This was when outsiders came in large numbers to the region mainly for commercial purposes. According to Sen (2011), colonial State policies regarding this region allowed the exploitative outsider to use this region for commercial purposes, changing the demography and societal formation of the Ho Adivasis. The colonial State moved from autonomy to taxation that organized the relation of production in this region (p. 21). This move of the colonial State can be conceptualized as a shift from 'a regime of sovereign power to a regime of governmental power' (Gidwani, 2008, p. 7). Gradually, the colonial State introduced the Land Acquisition Act (1894) to facilitate the market economy through primitive accumulation. At the same time, the land tenure legislature, CNTA 1908, was introduced to address the mass protest and uprising. This act restricted the land transfer and sale of tribal land to non-Adivasis in 16 of Jharkhand's 24 districts in the central and western region. The State even manipulates it as the name of the mining as synonymous with development in this region (George, 2009; Kumar & Mishra, 2017).

Not only land is encroached on and leased out, but the forest was also regulated through State laws and policies from the colonial times. It significantly

Table 4.2 Kolhan Policy Changes from Pre- to Post-Independence

Year	*Nature of Policy Changes*
1837	Formation of Kolhan Government Estate; Ho-dominated Singhbhum placed under the direct administration of company's government
1878	The Indian Forest Act 1878 had created a reserve forest, protected forest. It effectively excluded the local Hos and other indigenous users from accessing the forest. That was followed by the general forest policy of the Government of India in 1894. Forest was divided into four groups: reserve forest, protected forest, village and private forest, and wastes. An era of preservation and conservation of forest had started.
1894	Land Acquisition Act
1908	CNT Act
1927	Forest Act and territorialization
1965	Bihar Land Bill
2000	Formation of Jharkhand

Source: Author

impacted the Ho tribe, which is closely associated with the natural forest. Forests served as various sites for them. The natural access of forest for firewood collection, bamboo, Sal leaves, fodder grass, thatching grass, and other minor forest products was restricted after the successive colonial forest policies.

British law was distinct from customary laws of Ho Adivasi. Leadership remained predominantly in the hand of village *Pir*. The colonial State preferred for the *Mankis* and *Mundas* to settle the inter or intra-village disputes; at the same time, they were used as a State agent to interact with the local community.

Singhbhum witnessed new land tenure and rights rules after the Kol rebellion in 1832. Kolhan was made into a government estate with a direct administration system created 'to keep the Kolhan as a reserve for the Hos and to preserve their communal social system'. The superintendent of Kolhan administered the area through the *Mundas* and *Mankis* of Hos Pir. In Singhbhum, 'three types of gairmazarua land were recorded: Gm Malik: which included waste and jungle lands, rivers, government bandhs, and major roads, Gm Am: including village roads, public tanks, sacred groves, and Gm Makan which included houses of non-agriculturists'. It markedly shifted from status to contract concerning property (Cohn, 1961). Nothing was permanent. All relations and systems of property and production were governed through a legal framework of a contractual system. Once broken, these laws have had an impact on State interest and were treated as a 'breach of laws' and termed 'a matter of the judiciary'. In the forest, the land was not naturally accessible co-equally. In the pre-State system, the notion of 'coequal access' (Jodha, 1990) was in trend; but it gradually started declining in the

State system. By this, he meant that all people of a 'village, tribe or community have equal access to non-exclusive resources'. Access to natural resources was influenced by factors such as *killi* and custom. In West Singhbhum, it was determined by which *killi* was dominant in the village and who had settled the village after clearing the forest. As the State process continued with the project of territorialization and other agendas, it could be observed how the State produced a politically defined space that determined the allocation and realization of resource access rights (Peluso & Vandergeest, 2011). Scholars have shown how a process of territorialization helps the State to convert natural territory into a civil administrative unit. Through that, the State took over the administration of the right to land and forest where this project has been completed or proposed to become. The rule of law and the mechanism of this process determine the allocation and realization of resource access rights. Additionally, it also controls the self of the people and the community who use such resources (Peluso & Vandergeest, 2011; Peluso & Watts, 2001).

Procedural and eventual changes among the Hos after the State formation in colonial times showed how territorialization had disturbed the socio-ecological life of the Ho Adivasi. The Hos of Kolhan were a martial race before they were subjugated by the British in the first quarter of the 19th century. Their habitat was a sealed territory to outsiders, and they developed their culture on the forest-clad plateau of *Kolhan* undisturbed by outside influences. The conquest of the British brought in its train the much-hated *dikus* (foreigners), who immediately began economically exploiting the people and their habitat. The colonial power soon destroyed the martial spirit, and they seized the region. The population increased beyond the supporting capacity of the land resulting in the growth of a class of land of landless labourers who now flock to the mines and factories which the *diku* has established in the heart of the *Kolhan*. They came in contact with outsiders of questionable characters. The establishment of administrative towns, market towns, schools, and hospitals, and the opening up of railways have also contributed to colonial State formation. The social organization, material culture, institutions, and economic life of the tribe were undergoing profound changes; most of the intervention came for the betterment of the Adivasi. Abandonment of the villages, the break-up of moral life, division of the Adivasi into two communities, or many segmentary groups in the pattern of the Hindu caste system, growth of a class of landless labourers, adaptation of money economy in place of barter, use of luxury articles purchased with staple food, taboos on healthy food in imitation of the Hindus, and the development of parochial consciousness in the political field are some of the evil influences of the contact with the advanced (Chatterjee & Das, 1927). This was not only the case with the Hos but can also be found in other Adivasis where they have come under the direct influence of advanced people.

In Bengal, Bihar, and Orissa, the East India Company became the revenue collecting agent (diwan) of the Mughal Emperor on 12 August 1765.

However, the British had to engage in numerous tribal revolts throughout their rule (Areeparampil, 2002, p. 55). It led to colonial repression, inflicting casualties on both sides but ultimately resulting in Adivasis' subjugation. Paying fixed revenue to an external power instead of some arbitrary tribute to the headmen reaped apathy among the Adivasi. Moreover, the British also established their control over forests disregarding the customary rights enjoyed by the Adivasi (Gadgil & Guha, 1992; Gadgil & Guha, 2013). Thus, the colonial revenue policy and the forest laws, which came into existence after a prolonged debate over 'rights' versus 'privileges' during the 1870s, were the main reasons behind their discontent (Gadgil & Guha, 1992). This step also alienated the tribals from their earlier resource base and provoked the (mis)use of forests by the Adivasi, mainly against their cultural norms (Guha, 1989). Hence, the alienation was dual in nature and came from the material and cultural outcomes.

The colonial authorities convened massive deforestation for the construction of railways. Even though the loss of the Adivasi land is assumed to have started in the 17th century (Weiner, 1978, p. 158), 'the colonial policy of deforestation had the underlying intent of converting forest land into degraded land. Such land was used to generate revenue by granting agricultural usage rights, mainly to outsiders'. In fact, it was under British control that the 'non-tribals immigrated in huge proportions into the Adivasi territory' (Weiner, 1978, p. 161). They established their control over businesses, educational institutions, administration, courts, and the almost entire life of the Adivasi. These policies resulted in deplorable living conditions for the Adivasi. Their responses were articulated in the form of rebellion by Tilka Manjhi (1784), the Ho revolt (1820), the Kol insurrection (1832), the Santhal rebellion (1855), and the Birsa movement (1900).

The *scheduled areas and scheduled tribe commission*, popularly known as Dhebar Commission, in its report submitted to the central government of India in the year of 1960–1961 argued that the tribal, who formerly regarded himself as a lord of the forest, was through a deliberate process turned into a subject and placed under the forest department. Tribal villages were no longer an essential part of the forests but were there merely on sufferance. The traditional rights of the tribe were no longer recognized as rights. In 1894, they became 'right and privileges', and in 1952 they became 'rights and concession'. Now they are regarded as a concession (Dhebar, 1961, pp. 129–130).

The report submitted by the Dhebar Commission correctly mentioned the facts about tribal marginalization concerning State laws and policy. The Forest Act 2006 was an attempt to provide recognition of the right to access and assert their control through legal processes and steps prescribed by the State. The act proclaims that it resulted from a protracted struggle by our country's marginal and tribal communities to assert their rights over the forest land they traditionally depended on.

4.5 Emerging Political Identity and State Formation

The material and moral marginalization of the Adivasis provoked them to conceptualize new forms of political participation, usually known as *Adivasi Swashashan*, i.e., tribal self-rule. This form of governance was based on the ideology of '*Humara gaon mein humara raj*' (our rule in our village). A sense of ethnicity developed, and the Adivasis regarded themselves as 'we' and the outsiders as *dikus*. The newly formed Adivasi identity revolved around Adivasi dishum, which is 'our land, Adivasi land', and political mobilization began around the agenda of control over their land and life. Against this background, 'the pre-independence socio-political mobilization of Adivasi can be divided into three major phases' (Singh, 1982, p. 2). The first phase started at the turn of the 20th century and continued till 1920.

The second phase lasted from 1920 to 1938. An Adivasi urban middle class was taking shape, strengthening the inter-denominational unity for political purposes in a sustained manner. Despite several attempts to spread its activities to villages, it remained limited in the urban centres and can be termed as an 'urban movement'. This period also witnessed the development of cultural revivalism among the Adivasis, which stressed the return to the golden age of the community (Abua Raj or Birsa Raj) and formulated a religious sectarianism approach to reform the Adivasi society. The revival of many customary institutions like *Parha* (ancient Adivasi assemblies), Manki Munda Sangh, Adivasi Ho Mahasabha, and the discovery of lost 'scripts' are said to have occurred during this phase (Singh, 1982). Amidst all this, one noticeable event took place before Independence when Jai Pal Singh Munda, a formidable Adivasi leader, displeased with the top leadership of the ruling Congress party demanded recognition and separate homeland for the Adivasis.

The third phase, from 1938 to 1947, saw the rise of a militant movement under the Adivasi Mahasabha. The electoral debacle of the Adivasi in 1937 in the hands of Congress compelled the Adivasi groups to join hands for the formation of Adivasi Mahasabha. This Adivasi organization had a broader base and possessed pan-tribal composition and visions. The Jharkhandi movement, Bengali–Bihari (insider and outsider) controversy, and trade union politics were the other factors that accelerated this movement after independence. After the third phase, the demand for a separate State movement known as Jharkhandi movement was continued against the Bihar State's internal colonialism (Corbridge, 1988, 1993b). This movement also geared up when two different political parties Jharkhand Mukti Morcha (JMM) and the Communist Party of India (Marxist) participated in Adivasi negotiations with post-colonial State structures over regional autonomy (Chandra, 2015b) further extended by other regional political parties like All Jharkhand Student Union (AJSU).

A blend of conflict and conformity characterizes Adivasi movements, protests, and demands as they seek to attain their goals in the negotiation domain

(Chandra, 2015a, 2015b). This negotiation gives them a sense of identity that is confined to a particular territory. Apart from these three phases, providing the prize to address unrest and cope with the local traditional chiefdom with the modern State formation process continues. It was the negotiation of the State from the below.

In colonial times, the treatment of Munda Adivasis different from that of Hos was also one of the examples that confronted the identity of Adivasis, and their reactions against the State and its developmental agendas have also labelled Adivasis to describe themselves with broader and narrow political issues imbued with their sociopolitical aspiration. Here it is necessary to mention that State negotiations can be different for different tribes. Still, State also gave a pan-India collective identity when it amended any law related to land and resources of scheduled areas (Chandra, 2013b, 2015a). Chandra mentioned that the efforts of State imply that one is no longer only a Munda, Kondh, or Bhil but a member of an emergent all-India collective. Singhbhum of Jharkhand represents a different picture when Adivasis claim that they are the natural inhabitant of the region. The democratic politics and State formation in the Adivasi region also consider their claim. It has shown their claims and assertion and ability to negotiate with the State to some extent and articulate a different political vocabulary beyond the concept of tribe.

4.6 Post-Colonial State and Production of Marginality

Post-colonial State developmental policies and acts are the same as colonial State, but control of territory and territorialization projects are still vital for the State. Due to the decisive role of the State in regulating economic affairs, the power and control over natural resources through policies and acts are critical for the continuity of the State–Neoliberal agendas. Developmental discourse and projects are crucial to the State because they represent possibilities to materialize and politicize the national interests for development through processes of State territorialization (Vandergeest & Peluso, 1995). To favour the capitalist class and industries, the State employs distinctive strategies to regulate socio-spatial relations within a particular space. These strategies create differentiated meanings of territory, define new regimes of land use, and finally legitimize emerging economic activities (Agnew, 2013; Sack, 1986). It also has a substantial impact on tribal identity and landscape. A report of the committee on forests and tribals under the chairmanship of Prof. B.K. Roy Burman stated the agony of the Adivasis. The report said:

> Tribal have evolved a way of life which, "on the one hand is woven round forest ecology and forest resources and on the other ensures that the forest is protected against the depredation by man and nature.
>
> (Government of India 1982, p. 11)

The territorialization projects of the State not only degraded the 'landscape'[22] of Singhbhum but also led to the proletarianization of the Adivasis (Areeparampil, 1993). One factor that led to proletarianization was the loss of the symbiotic relationship between the Adivasis and their ecology. Another report by the Government of India (1983) also mentioned that the symbiotic relationship suffered a setback during the colonial rule when the forest was viewed only as a source of maximization of profit and not a vital link between the human habitat and the larger environment (Kumar, 1983). Rapid deforestation due to extractive industrialization made the forest department and authorities more concerned about damage control with the rate of financial return and net revenue, shifting the forest department's engagement with the Ho community. Afforestation, forest diversion, and commercial tree plantation were such initiatives that showed the concern of the forest department. Saranda (Asia's largest Sal tree ranges), where earlier sunlight was unable to touch the ground, is now half deforested due to surface mining. It has induced marginalization of the local people who were dependent on land and forest for their livelihood. The 'marginal' are those who have been either forced off the land or forced on to poor or insufficient land (Li, 2005b). They often live in the most dangerous and unhealthy places. They are the worst sufferers of growth. It is believed that industrialization and capital penetration have induced a division of labour within the communities and society, and market forces have forced them to be the poorest of the poor, to live in the place that was their territory and rich with varied resources.

Their marginalization became a subject of State description that has been continuously defined at the centre from colonial times onwards. The changes and transformation of the ecology are intimately linked with the control and exploitation of indigenous resources by the State to favour the capitalist class. The underdevelopment process forces the peasantry into a more vulnerable position, which, in turn, 'directs them to look for another source of livelihood in areas where security may be less and hazard more severe or to change their resource use in ways that exacerbate vulnerability' (Susan et al., 1983, pp. 277–278).

As discussed earlier, territorialization, State exercise of power, and developmental agendas were designed to accomplish State formation. Modern State formation began from colonial rule with the process of territorialization. Through land settlement, imposition of bureaucratic rule, and institutions for village administration, the State has reached the local people. Corbridge, Williams, Srivastava, Veron, and Véron (2005) have questioned as to when and why the local people of this region should interact with the State. They said,

> 'as soon as we begin to think about the when and why of the poorer people's encounter with the State, we run into problems. These encounters will vary significantly from place to place, and over time. They also depend

> on which agency, or "who" in the State is being seen. In the remote interior villages of the Singhbhum District, Bihar (now Jharkhand), it was common in the late 1970s for forest-dependent villagers to see the "State mainly in the person of a forest guard, or a panchayatsewak.
>
> (p. 21)

A panchayatsewak worked as a State agent and served as a bridge between the State and the local community. As the panchayat system strengthened in the form of Panchayati Raj, the traditional form of Panch (village elders) started to degrade (Puthumattathil, 2014; Sen, 2018). People's interactions and encounters with the State happened through the State agencies and institutions. Their identity is located exclusively in Scheduled V areas under the Indian Constitution. Here it is necessary to mention that the State was not so alien for the Adivasis or after colonial State formation; the State was very much present through its agencies, which continue till today but in a very reductionist and shadowy form. It can be understood through the history of Scheduled areas. From the colonial times, the provision of the backward tract, the Scheduled district, recognized its intimate form in the post-colonial constitution in the form of Schedule V. Modern State-making emerges out of the intersection of colonial forms of knowledge (Cohn, 1996) and specific characteristics of post-colonial rationalities. State rationalities also shape Adivasi imagination and politics. Adivasi politics vis-à-vis the post-colonial Indian State is characterized by negotiating and periodical claims of the Adivasis to State that deal with the issues of development-induced displacement, dispossession, marginalization, and future insecurity. These issues better showed how State–Adivasi relations in legally demarcated Schedule V areas are in trend.

Top-down accumulative developmental projects fail historically to deliver their promises. The development process is seen to come from above that conceives West Singhbhum as traditional, static, economically backward, and underdeveloped. This kind of project degraded the local ecology and is still in practice. It is hard to imagine a situation where half of the land and forest area are leased out, accumulated, or ready to be leased out for mining companies and developmental projects without the knowledge of the Ho people.

These top-down policies and projects come through bureaucratic channels and integrate into the local level without consideration of the pluralities of the relation between the Adivasis and their local ecology. During field work, an incident was reported in Manoharpur block, where State officials had come to demarcate the boundary for the mining companies, which the local villagers opposed. Continuous resistance forced Vedanta, a mining company, to shift their mining plan from West Singhbhum to another district of Jharkhand. The government has to take mining decisions following large protests against the State accumulation policies. Many incidents have been reported which show how the local villagers oppose State developmental policies and projects and, afterwards, how the State starts to negotiate at the agency level through local governmental officials. This opposition and negotiation also produced environmental identity and objects. Collective protest and resistance gave them a representation of self-identity derived from historical myth. The region has a history of resistance against the State for getting 'swaraj' and autonomy. Still, development does not give them liberty and 'swaraj', but has created a sense of fear in the form of displacement and dispossession.

Mass protest rally was organized against the State amendment in land acquisitions and discriminatory policies near *Saheedudhyan* (Martyr garden) in Chaibasa, which is an administratively and politically influential region (headquarter of the Kolhan region). This shows how local people are in fear for their existence and survival. In his speech one of the local leaders said, 'we are here to fight against the State amendment of CNT act[23] (*Aar Paar ki Ladai*)'.[24] The State tries to dilute the Chotanagpur Act's provision, which directly impacts our society. It will destroy our identity. A few days later, Johar, an active human rights-based organization, helped to create an opportunity to interact with D.N. Champiaji (ex-member of Parliament and ex-speaker of United Bihar Legislative Assembly). During our conversation, he expressed his agony as follows:

> State amendment of CNT was very unfortunate, but it is the politics to attract companies. Our source of livelihood has declined over the period. Our dependence on nature and ecology is not as before. If it is passed in the assembly and comes in the form of a law, we will no longer be able to save our 'Adivasiyat'.[25] Jharkhand was made for Adivasi, but now it is for the outsiders and corporates (*dikus*). The livelihood sources and economy are no longer subsistence and productive. Adivasi are afraid because they don't know when their land will be leased out to industrialists.
>
> (Personal interview dated 20 November 2016)

His narration and fear for the land leased out were really worrying and true in the context of the Ho community of this district. This also shows the State's treatment of its people and natural resources that requires detailed description. The tribal protests and resistance against developmental politics show a contradictory scene and experience of development at an agency level. The top-down development projects of the State in the form of the extractive industry have been protested at a mass level. It is a fact that developmental projects in the form of resource extraction became a curse for them; this requires a discussion on a broader frame of State formation which will expose the everyday activity of the State. State and Adivasi communities' relationship within the process of post-colonial State formation is based on negotiations. These negotiations with the State are also negotiations of the

State power, which can be observed when Adivasis protested against the State and the State returned with a bundle of approaches, including a populist scheme, *Dhaba* construction, subsidized ration, and other things. To this extent, the negotiation of the State at agency level and the Adivasi protests and movements are characterized by the blend of conflict and conformity as they seek to attain their objectives. Their relationship in contemporary India still follows the pattern of negotiation and accommodation. Social movements, protests, and claims to assert their identity are the principal mechanisms through which negotiations with the State occur.

4.7 Linking the Narratives of the Changes due to State Formation with Five Theses of Political Ecology

The discussion on the history of State formation has shown some conceptual narration in the form of territorialization (conservation and commodification), marginalization, exclusion, identity, and resistance. These can be linked to the five theses of political ecology as discussed in Section 4.2 of this chapter. The issues that come also offer practical clarity to the theses of political ecology and help to locate our discussion in the context of West Singhbhum. It provides a conceptual foundation for the Adivasi–State relationship. It helps us to understand how specific politics of State in the form of territorialization, conservation, negotiation, and other developmental projects are a reason for the transformation of the ecology. These issues are interrelated and not demarcated into a singular frame. The uses of the five dominant theses of political ecology showed how each issue and narrative of development is related to State formation, and each issue and thesis of political ecology is linked with other narratives (see Table 4.3). It all contributes to building a complete panorama of environmental or ecological changes of West Singhbhum.

Narration one, along with its statement '*nadi nale sab sukh gaye hai*', explains the over-exploitation and misuse of the resources by the capitalists and mining player, whereas in a supporting argument Ramesh Jerai, a human right activist, mentions how State-facilitated market penetration has been increasing, which forcibly shifts our system into a profit-based capitalist system. These two narrations on degradation and marginalization focused on the critical link between the ecology and the resource. It also showed the State developmental programme in the form of mining, deforestation, and territorialization resulting in resource degradation and scarcity. It also challenges the notion that State formation and its developmental agendas in the form of neoliberal policies and extractive capitalism can lead to the wellness of local people and ecology. Instead, the history of State formation among the Adivasis reveals that the development programmes of State are programmes that fail. The State's primary concern is facilitating capital-centric growth, while its top-down policies and projects have failed to deliver on their promises of tribal self-rule and autonomy. State territorialization and

Table 4.3 Examining Narrations through Five Theses of Political Ecology

Narrations	*Theses of Political Ecology*	*State Formation and Changes in West Singhbhum*
Nadi nale sab sukh gaye hai [English trans: River and water bodies are dying]	Degradation and marginalization	Ecological degradation, long blamed on local people is shown in its larger political and economic context, territorialization, and over-exploitation of the resource to fulfil the demand of global market
What do we villagers do if all land are leased out to extractive companies in the name of development and public purposes	Conservation and control	Protected forest and reserve forest declared from colonial time but for mining forest diversion or deforestation can be done.
We do not want Mittal, Jindal, and Vedanta. We have to remember our past and teachings of Birsa Bhagwan. If we do not protest, nothing will leave. Birsa Bhagwan taught us *Abua Raj Seterjana, Maharani Raj Tundujana* (Let the kingdom of queen be ended and our kingdom be established). After getting independence, we are in the dream of *Abua Disum Abua Raj*' (Our Village Our Government)	Environmental conflict and exclusion	Environmental conflicts are shown to be a part of larger class struggles and vice versa. Adivasi protest against capitalist class shows the class formation among the Adivasis.
We have no job; our diversified source of livelihood is gradually limited	Environmental subjects and identity movement	Political identities and social struggles are shown to be linked to be basic issues of livelihood and environmental activity, Jharkhandis and Abua Raj (self-rule) are the new subjects
Na Vidhansabha Na Rajyasabha Sabse Bada Gramsabha (neither State Assembly nor Rajya Sabha (Upper House) are bigger than Gram Sabha (Village Assembly))	Political objects and actors	Local-level civil society emerged against the State that encourages the peasants and tribals to fight against the State's neoliberal policies

Source: Author

conservation politics jointly have an impact on the ecology of Ho. They are based on the State's coercive power, which is dispersed into all agencies and institutions of the State. It enables the State to exclude and include local people in the form of spatial fixation and mapping in the name of conservation and territorialization. It produces cultural effects in the form of marginality (Tsing, 1993). Therefore, an understanding of marginal people and the factors responsible for their marginalization is best understood through the State hegemonic project. These projects have the disciplining effect that is always a subject of the State.

Ideological apparatus: State-produced marginalization is also a social construction that involves a process through which particular communities and populations become a subject of State periodic and regular description. Narration 2 is as follows: 'What do we villagers do if all land will be leased out to extractive companies in the name of development and public purpose'. Considering the region of West Singhbhum as a socio-ecological system, it can be said that the Adivasis were blamed and labelled as backward and always in need of intervention. Still, the projects that came to civilize them resulted in deforestation and ecological degradation. The State's conservation policies became a techno-sphere that has been tactfully used only to give clearance for mining. Once made to protect the region's forest and community wellness, the forest department is now limited to favour State's neoliberal agendas. From colonial times till now, the State has controlled the forests to achieve better conservation, but it is a fact that forest ecosystems are dying and rapidly degrading. A report submitted by Justice Shah in 2013 mentions how illegal mining and deforestation have taken place in this region. The State only paradoxically liquidates the money in the name of tree plantation, which is also not culturally significant. The State first took control of the region and forests through several forest acts and land tenure bills. This showed how State formation controls the territories and gives several meanings to these territories in the form of land and forest classification. Community-based systems have evolved over a historical time, and the local people interact with nature through their culture and customs. Their practices and customs are completely overlooked and ignored.

Third narration, 'environmental conflict and exclusion', deals with the issue of resource access, distribution, exclusion, and control. It shows how forest acts, land acquisition acts, and mining projects of State problematized the ecology that induced conflict. The co-equal access of resource access is disturbed. It raises several questions about equity, justice, and participation in a developmental project. It reveals that State arrangement as the techno-political development project is exclusive, exploiting the local ecological users. Class conflict is based on different philosophical foundations and ideologies of the use of natural resources. State facilitating a neoliberal capitalist class exploited the local resources that induced conflicts and resistance. It also developed class dynamics within the homogeneous class of Adivasi that promote heterogeneity. Those who oppose the State developmental projects in the extractive industry belong to defenders of their moral or peasant ecology.

The fourth narration deals with how environment identity is newly formed and produced. Local people are labelled as marginal, displaced, and backward, a new identity that came after State intervention and formation restructured and reformed Adivasis regularly. State marked Adivasis and their territories regularly as a routine practice of hegemony. Making of Jharkhand, or '*Jharkhand Jharkhandi ke liye*', was such a constructed reality that gave a new identity to Adivasi as a citizen of Jharkhand. The discourse and the movement for Jharkhand were related to ethnic conflicts and livelihood issues. Still, after Jharkhand's formation, industrialization was a primary path that betrayed the hope and aspiration of the Adivasi.

Fifth narration, 'political objects and actors', deals with how capitalism as a sociopolitical condition operationalized in the extractive industry affected the landscape and ecology. It also shows how the interaction between two sociopolitical systems have resulted in transformation and changes. The history of the State formation of Singhbhum showed that different agents and classes had developed new roles and networks based on their interests. One of the major interpretations may be further explored in this connection: how hegemonic institutions and different corporate companies along with forest officials directed and manipulated the power to favour mining companies that led to unintended consequences. This process also induced everyday forms of resistance (Scott, 1989) that emerged from traditional, alternative, or progressive human groups (cutting line of ethnicity, gender, territory, and country). Protests against State and the claim that '*...sabse bada gram sabha*'[26] are also an attempt to regain their identity through negotiation with the State from below. It also challenges the top-down State scheme and developmental projects.

These theses of political ecology seek to explain why environmental systems change (because of State formation and State facilitation to industries, integration of local ecology into market or capital accumulation). It explains why people resist developmental projects (due to capitalism and State policies which promote neoliberal developmental projects and activities) and why social identity changes (because of transformed environmental institutions). The States in Adivasi regions are experienced in a pluralistic sense. Therefore, it requires a systematic discussion that engages the State with nature, culture, and ecology.

4.8 Conclusion

Since colonial times, resource-rich regions have been defined, organized, controlled, subjugated, and restructured through various developmental policies and practices. State interventions to develop these regions are primarily economic and political, serving particular agendas. To improve the region and the community, the State has been exercising power through developmental projects and schemes that Foucault acknowledged as 'governmentality'. Li (2007) argued that multiple tactics, calculations, and finalities are prioritized before 'development' interventions are initiated. These developmental interventions are traversed by political will having multiple intentions, developmental, visionary vocabularies, techniques, devices etc., and have worked for the transformation of the society. She mentioned that two practices 'problematization'[27] and 'rendering technical'[28] are required to translate the will to improve into a developmental programme. These two practices relevant to developed nations provided a capacity for the State to create and diagnose the deficiencies through a calculated intervention programme. It helps the State to operate and adjust in a very corrective manner. In developing nations, it is all coercive as essence and idealizing the remedial measures as the appearance of development.

Corrective measure in the representative form of the verb 'to improve' is meant to turn agricultural land into a profitable commodity (Drayton, 1982; Williams, 1976). Williams has given an example of medieval England to illustrate the verb 'to improve' to turn agricultural land, common land, grazing land, and waste land into a profit land with surplus value. The limits of each government from colonial State to post-colonial State formation show how failures and deficiencies are continuously reshaping the ecology of Jharkhand. The continuous shaping and treatment or the loss of developmental intervention is a new opportunity for new interventions to correct deficiencies (Li, 2007). As the intervention is deepened through policies and developmental projects, transforming the ecology and local production system will continue rapidly.

This chapter has explained the transformation of the State and society through the political–ecological narration. It shows territorialization, conservation, and commodification-induced marginalization, which is the sum of unequal social situations that reduce options for life (Blaikie & Brookfield, 1987; Li, 2014; Robbins, 2012). This chapter has also elaborated the five theses of political ecology to deal with the issues of changes in West Singhbhum. It is observed that State's development narrative argues that resource extractive industries are for the betterment of the local people – the official account claims that with the production of minerals life choices of Adivasi have been increasing – but paradoxically the Hos experience that State projects have adversely affected their ecology, culture, and livelihood. It can be said that the State exercises its power and control but never guarantees the beneficial outcome and its promised delivery. Adivasi region of Jharkhand has a history of continuous failed improvement schemes and diagnosis. The planned extractive industrial activities of the State have failed to deliver their promises, and the diagnosis has been incomplete, but it has routinely created more sense of marginalization from the colonial times.

Chapter 5 makes a descriptive analysis of developmental projects, resistances, and the issues of governance that shaped the State–Adivasi relationships. It also discusses how State development programmes facilitate capitalist forces, which induce everyday resistance and contribute to the systematic weakening of the local Adivasis' moral- and subsistence-based economy.

Notes

1 See Annual report 2004–2005, published by the Department of Food, Civil Supplies and Commerce, Government of Jharkhand pp. 50.
2 See mining map published by Government of Jharkhand in 2011.
3 See Jharkhand, the mining base of India, published by IBEF. https://www.ibef.org/download/Jharkhand-March-2017.pdf.
4 https://www.ibef.org/download/Jharkhand-March-2021.pdf.
5 The depletion in groundwater level is reported in the Hindi newspaper *Daily Prabhat Khabar* (3 April 2016). According to the report: West Singhbhum District experienced scarcity of drinking water and drought also made the situation pathetic. The district has 11,250 hand pumps for drinking water and other uses of which 2,548 are defective due to the depletion of water level; many pumps are useless without providing any water. Groundwater is also alkaline in nature. In plain areas, due to the mining of graphite, limestone, and other minerals, collapsing of borewells is also reported. People of the hilly region depend upon the dirty water of river streams and natural tanks (chua). Kolhan has 22 rivers and West Singhbhum 12. During the field study, it was found that nine rivers are totally dry and three rivers have polluted water. Now the dry river is used for sand mining, and the river with water in West Singhbhum is polluted with red oxide and mining wastage.
6 In West Singhbhum, 13 lapse are authorized to sell forest products. The state also institutionalized the forest products through minimum supporting value and other initiatives. It enhances the State's ability to get benefit from the product, but it also creates a problem for the Ho Adivasis to access these lapses to get the minimum supporting value of the forest products.
7 To understand why people are worried for CNT amendment, see https://scroll.in/article/828802/in-jharkhand-Adivasis-say-changes-to-tenancy-laws-dilutes-their-hard-won-land-rights.
8 This narration was based on the meeting and personal conversation with the people of the civil society and activism. Conversation and interview with Veer Singh Sinku, Bhal Bhadra, D.N Champia, and Xavier Dias helped to brief the situation of West Singhbhum through the narration.
9 Smallwater tank.
10 Villagers of the Noamundi and Jagganathpur block protested against the State. They were demanding work. After the closure of mining, they also lost income sources through manual labour. They often protested and demanded that the company provide them with some work. An old villager residing in Noamundi Basti said that there is no work in this region. Mining has degraded the life choices and forced them to sell their labour at a cheaper rate.
11 Doboro Bullibuli, a Ho poet, activist, and educationalist, mentioned that their customary institution is gradually co-opted by the State through different ploys (Doboro: personal interview 4 February 2016).
12 John Miran Munda and his colleagues actively, through rallies, posters, and campaigns in the nearby region of mining-affected villages, demanded for the

autonomy of Gram Sabha. They mentioned that Gram Sabha is a supreme and basic unit of the people's democracy. In mining villages of Tonto, Jhinkpani, and Noamundi block, this poster is advertised in the muddy wall of villages. After investigation from the villagers, we found that some groups of the villagers opposed State intervention in the name of mining projects. They argue that through Gram Sabha they will decide what kind of developmental projects are suitable for the ecology of the region.

13 These narratives are seen as being an alternate explanation of the ecological change. It can be termed as 'received wisdom' (Leach & Mearns, 1996, p. 3). Narratives are a part of a broader analysis of the discourses concerning environmental issues, identifying key actors, interests, and issues. For more discussion, see (Nunan, 2015, p. 33).

14 State's understanding of wilderness is in the form of nature without people.

15 Ecologize means how the State creates a procedure to develop a network of quasi objects in which relations of subordination remain uncertain. For more discussion, see (Latour, 1998), http://bruno-latour.fr/sites/default/files/73-7TH-CITY-GB.pdf.

16 Capitalism as a dominant ideology subsumes all ideologies, and dispossession can be aligned with a widely accepted notion of national development. Where the capitalist ideology and State ideological and monetary justification prove inadequate, there resistance continues, but the State by vices and virtues, by using its ideological and repressive apparatus, is able to balance the lineage of the resistance and other issues.

17 Resource nationalism is defined as the consideration of natural resources as part of the national heritage whereby the management and use of resources are aligned to national priorities.

18 For more discussion on how territorialization and commodification is used by state, see http://territorialmasquerades.net/territorialization/.

19 It denotes the people who clear the jungle to settle the village; for more, see Singh (1978).

20 Imperialism is the highest stage of capitalism. For more discussion on imperialism, see Lenin (1999), https://www.readingfromtheleft.com/Books/Classics/Lenin Imperialism.pdf.

21 The mode of production is 'a definite forms of activity, of those individuals a definite form of expressing their life, a definite mode of life on their part'. As individuals express their life so they are what they are, therefore, coincides with their production, with both 'what' they produce and 'how' they produce. The nature of individuals thus depends on the material conditions determining their production, in Karl Marx and Engels, *The German Ideology*, 1968, Moscow p. 32.

22 Landscape is a medium not only for expressing value but also for expressing meaning. It is not only a communication between man to man or human to human but, most radically, it is for communication between human and non-human. For more discussion, see (Mitchell 1994. *Landscape and power*. Chicago: University of Chicago Press).

23 Land of *Adivasi* was non-transferable under CNT Act for commercial purpose.

24 Final war and negotiation.

25 A way of prosperous life under the lap of nature (*Jal, Jungle*, and *Jameen* are the three important elements for this philosophical foundation of *Adivasiyat*).

26 Village assembly is the supreme one.

27 Problematization is identifying deficiencies that need to be rectified. It is used to identify the problem. The identification of problem is linked to availability of solution and this all is merged within the government assemblage.

28 Rendering technical are those who are in position to diagnose the problem.

Chapter 5

State-Stimulated Extractivism, Accumulation, and Resistance

And what are the sacrifices the State makes to the colonialist, the darling gods and exporters? The answer is simple: it sacrifices the property of the natives to him. ... The natural produce of the colonized country grows on the land and this land belongs to the 'indigenous' population. In certain thinly populated regions, with large uncultivated areas, the theft of land is less apparent: what you see is military occupation, and forced labour. (Sartre, 2005, pp. 11–12)

5.1 Introduction

State capitalism and people's sacrifice are the histories of revolutionalizing nature. Its distinctiveness lies in the facilitation of endless accumulation (Moore, 2005). Indeed, it does not have an ecological regime. It organized itself in a quasi-stable relationship between humans and the rest of nature that facilitates the endless accumulation of resources through the political economy of the State. Natural resource is a significant determinant of the State's capacities which also determine the region's development (Mann, 1984; Tilly, McAdam, & Tarrow, 2001). Territorialization and commodification of the natural resource became the State's primary means for implementing its extractive projects that go with the State-building programme. By attracting investments in the resource sector, the State promoting the act of extraction has worked for the spatial movement of the capital. The process of financialization, economic privatization, and capital accumulation promoted by the State through its legislative policies and executive political plans has led to what (Gudynas, 2010) refers to as 'extractivism'. The State does it to achieve the promise of perpetual growth, which can tune with the creative destruction of neoliberalism. Like any other social site, the State is overdetermined by all the socio-ecological processes. The capitalist State in trialectics provides certain natural, cultural, political, and economic conditions unique to each class structure that favours extractive capitalists. Land-grabbing wars, development-induced dispossession, and ecological resistance have taken a central position in the political economy of India.

DOI: 10.4324/9781032637990-5

It has led to a high level of attraction and interest for scholarly works throughout the globe. This chapter argues that State's character in these land deals is based on the accumulation and appropriation of resources that treat nature as a cheap raw material. The resulted resistance is multifaceted and varied, which questions the character of the State that negotiates and deals with these populations differently.

This chapter aims to understand what resources are meant for the Hos. Why have they been struggling and resisting against the mining companies and developmental plans of the State to save their ecology? It also shows how the State's facilitation of extractivism and its illegalities are based on power asymmetry that impacts Ho territory. Based on arguments in the previous chapters regarding the State formation and changes in terms of dispossession and marginalization, this chapter tries to uncover the State character and its illegalities. It also tries to answer some critical questions like causes and strategies of resistance, what locals got from mining industries, and what urged States to favour mining companies. The chapter employs varieties of concepts ranging from primitive accumulation to co-optation. These concepts are the tools of the political ecology perspective that help to explore the developmental outcomes of environmental and resource politics in terms of resistance, assimilation, and injustice.

State developmental strategies have appropriated nature and resources and given it in the hands of extractive capitalist players that induce dislocation and dispossession of Adivasis from their territory and means of production. It has also resulted in everyday resistance and everyday negotiation at the agency level. In addition, I will discuss several cases related to the State's facilitation and manipulation of laws to favour capitalist players and the ways and factors that led local Adivasis to resist. The Hos areas are acting as barriers to collective actions and failure of long-term movement.

In India, the State's top-down policy of development was a kind of democratic socialism that could be observed in successive five-year national plans (Kapoor, 2012). After independence, it mainly focused on industrialization, large-scale power projects, dams, and infrastructural development. These accumulative plans and projects had a history of development-induced dispossession, displacement, and marginalization in the State-building process, which gradually destroyed the social and ecological condition of the region. The influential work of Karl Polanyi entitled 'The Great Transformation' deals with how the capitalist market has degraded its own socio-economic and environmental conditions (Sahoo & Kumar, 2017a). His writing provides an insight into the problem of economic development and the social and natural environment, which can be best understood as a subject of the ecological limits to capitalist development. After 1991, the shift of the political economy of development from the Nehruvian to the Neoliberal market model relocated the discourses and practices of development, which favour corporate sectors and appeal for immediate resource utilization. The economic model of the State essentially nurtures the commodity fetishism, growth, and mechanization of the land, labour, and market. The neoliberal policies are claimed to be based on the twin Smithian beliefs of the free market (favourable condition for extraction of resources) and non-interventionist State; the State intervenes actively to create 'favourable' conditions for the capital, which are based on the grammar of the market (commodity's supply-demand) that is presumably connected with the metaphor of nation's interests. It demands territorialization, the commodification of natural resources, and the 'production of nature' through treating human and natural resources as governed subjects. The treatment of humans and natural resources is built on power and domination that can be easily observed at the time when the State interacts and negotiates with the different sections of society through its policies and plans. The power and domination of the State is a kind of epistemological power – to facilitate capital shaping the individual customs, body, time, and labour according to a particular system (Faubion, 2000). It works by transforming people's time – their living time – into labour time. Its second function is to convert people's bodies into labour power. In fact, State capitalist projects, now guided by neoliberal policies and activities, penetrate much more deeply into society.

States in the contemporary world are part of a neoliberal regime with the embedded presumption that market forces will ensure better distribution of resources to favour capital-centric growth and investors. But several scholars (Bagchi, 1982; Escobar, 2011; Harvey, 2004; Vanaik, 1990) take such ideas to be elevated by the developed countries favouring the cosmopolitan bourgeoisie and accepted by the developing countries favouring the semi-peripheral bourgeoisie. Government's plans and official policy work for the 'problem of absorbing capital surplus' (Harvey, 2007).

The specialized knowledge backed by the authority of the State changes the balance of power between the locality and the State (Scott, 1998b). In India, after the liberalization era, structural changes in public sector industries (partly privatized) are heading towards the maximization of economic profit in its motive and neglecting the earlier aim for overall development promised only to compete with the private sector industries. It was also a kind of specialized knowledge that means developmental projects now will be closely linked to capitalist production relations. The neoliberal avatar of the modern Indian State uses its apparatus (ideological and repressive) to accumulate land and natural resources from farmers/tribals to increase private and public projects: industry, infrastructure, and real estate. It resulted in reactive uprisings of the peasants and tribals (owners of the land) popularly conceptualized as 'land wars' (Levien, 2015). The appropriation of resources and control of natural resources has significantly contributed to the nation's economy; but at the same time, the community, especially 'ecosystem people' (Gadgil & Guha, 1992) and Adivasi people, bear the cost in the form of degradation of natural resources (e.g., deforestation, water pollution, flooding, and biodiversity loss) and subsequent loss of their ecological base of livelihood only due to the excessive resource extraction in the name of development projects. The appropriation and accumulation of resources (Baviskar, 1997b; Nilsen & Roy, 2013; Padel, 2014) and the emergence of extractivism (Bryan, 2012; Engels & Dietz, 2017; Haesbaert, 2013) have been opposed and protested by the community that is not a proletariat but the settlers of the place (Adivasis). Their opposition can be strictly conceptualized as what Harvey calls 'accumulation by dispossession'. This system of development is situated against the backdrop of accumulation, appropriation, and treatment of the natural resources (commodification); the 'counter resistance and collective efforts' of Adivasis seek to protect the environmental and social conditions that safeguard their ecological base of livelihood.

Scott (1976) argues, in the Introduction to 'The Moral Economy of the Peasant', that two major transformations during the colonial period in Southeast Asia' served State formation to undermine the pre-existing social insurance patterns radically and violate the moral economy of the peasant. First was the particular cultural system shaped under capitalism and, second, the development of the modern State under colonial rule. In India, during colonial times, the transformation of land and labour (i.e., nature and human work) into commodities for sale have had the most profound impact. Control of land increasingly passed out of the hands of villagers; cultivators progressively lost free usufruct rights and became tenants or agrarian wage laborers; the fluctuations of an impersonal market increasingly gauged the value of what was produced.

Scott's account was enlightened by the Marxian notion of primitive accumulation and Polanyi's work of 'the great transformation'. This capitalist era, where the State captures labour and produces commodity fetishism, has been tuned with State formation. As discussed in the Chapter 4, State formation depended, for its emergence, legitimacy, endurance, and entrenchment, upon nature and society through territorialization, commodification, and other developmental intervention (DI). The control, commodification, and sovereign power to allocate resources to different stakeholders and players are tied to the conceptual discussion of the political ecology of the State (Ioris, 2015), which can be represented as 'everyday tyranny' for the Adivasis (Nilsen, 2010). Everyday tyranny of the State determines the citizenship of the ecological natives (Adivasis) in two ways: the first one is the State's involvement in the violation of citizen rights and the absence of a universal legal order; and the second one is the social status conditioned by the State with a sharp economic disparity and inherited social inequalities helping the State to monitor and govern the citizens (Jayal, 2000; Nilsen, 2013). This inequality is due to the institutional arrangement of the State, which favours capitalist players through different legislatures, policies, and acts. It deepens the process of primitive accumulation. The primitive accumulation in

Marxian sense is associated with the formation of the modern State and its regulatory disciplinary agencies. This disciplinary power of the State transforms and subsumes the complex cultural logic of social ecology to expand its material domination over local territory. The capitalist player needs the power of the State, and they use it to control over the resources and capital. It helps them to maximize their profit while it forces the local community to sell their labour for the developmental projects what Li (2014) calls a dull compulsion in the Marxian sense through the several discourses of providing employment, developing the infrastructure and health facilities etc. at the time of their expansion. Their resource and the ecological base are accumulated, which correspond with the capital valorization requirement. It also forces them to keep working on capitalist players who provided life choices at normal levels to create a new kind of financial dependency.

5.2 Conceptual Framework

The expansion of the extractivism and the State's promotion of the capitalist's projects through its policies and apparatus often affect the people who have a moral economy. They react in terms of movement, protest, and resistance. Those Adivasis who resist have a moral economy that can be put in various conceptual frames. There is a nonlinear link between resistance and social change, which depicts how people or community resists and also in the system to minimize the tyranny and disadvantage of the extractive projects. Their interest is visible through social movements, posters, and rallies to revolution without reforming the power structures. However, those who resist may also support the structure to survive and question the authority to find a better place within the State. They are more interested in 'working the system ... to their minimum disadvantage' (Hobsbawm, 1973, p. 16). Resisters have varied objectives; acknowledging these objectives is crucial to understanding their resistance. Likewise, resisters may alter the mode of resistance only to suit their circumstances the best. This resistance is due to the primitive accumulation that dispossesses them from their land, culture, and territory. These extractions have a character of primitive accumulation (Marx, 1977). Political ecology provides tools for thinking about resistance produced by the politics of access and control over natural resources. Its focus on the power asymmetry inherent in defining, controlling, and managing nature suggests an alternative way of viewing the link between the environment and political action (Peluso & Watts, 2001). First, Political ecology treats struggle and resistance over natural resources as a struggle in defence of livelihoods. It has cultural, economic, and ecological dimensions related to different discourses and worldviews regarding the use of nature and its resources (Escobar, 2006b), unequal distribution of resources (Biersack & Greenberg, 2006; Li, 2010; Robbins, 2012), and environmental transformation (Li, 2005b; Moore, 1993), respectively. The political and cultural significance articulates the values, interests, and actions. In this vein, it is essential to understand that in the community that resists developmental projects and extractivism, their livelihoods not only are economically driven and structured but also have cultural and political significance (Bebbington, 1999, 2012).

The material struggle is not separated from the ideological struggle in peasant politics. To resist a claim or 'an appropriation is to resist, as well, the justification and rationale behind that particular claim' (Scott, 1985, p. 297). As a pattern of acts, it is undertaken by individuals or collectives in a subordinated position to mitigate or deny the claims of powerful elites and the effects of State domination while advancing their agenda to survive. Scott suggests categorizing resistance into two chief forms: the public and the disguised resistance (Scott, 1985). Both forms of resistance are linked to three types of dominations: material, status, and ideological. These three types of domination give rise to six kinds of resistance. Resistance is found to be publicly announced and in the form of open revolts, petitions, demonstrations, land invasions, etc. against (1) material domination; (2) assertion of worth or desecration of status symbols against status domination; (3) counterideologies against ideological domination. Further, resistance exists in the disguised form (low profile, undisclosed, or 'infrapolitics') as everyday resistance (e.g. poaching, desertion, evasion, and foot-dragging); (4) direct resistance by disguised resisters against material domination; (5) hidden transcripts of anger or disguised discourses of dignity against status domination; or (6) dissident subcultures (e.g. millennial religion, myths of social banditry, and class heroes) against ideological domination.[1] The varied form of resistance can be understood in a plural sense.

Moreover, political ecology observes conflict through the concept of accumulation through dispossession (Harvey, 2004). In this vein, some of the other concepts, power asymmetry (Bryant & Bailey, 1997), talk about the actor's desire to control or access nature that induces resistance and assimilation. Escobar's (2006b) concept of assimilation argues that incorporating the local population into modern discourse, processes, and practices of development also induces cultural conflict and resistance (Escobar, 2006a, 2006b). Third, it considers the strategies of resistance that Scott (1989) refers to as everyday forms of resistance. Scott also terms it 'the art of not being governed' (Dove et al., 2011), which is similar to Gramsci's concept of counter-hegemony (Mann, 2009). The art of not being governed talks about the history of deliberate and reactive statelessness, whereas counter-hegemony means resistance to the agenda of the dominant hegemony (Hall et al., 1996). Non-allowance of access to the natural resources to people or communities with a subsistence economy by the State leads people to resist and negotiate. Their mode of resistance and negotiation with the powerful State can be termed as 'weapons of the weak' (Scott, 1985). They adopt this method when they may be relatively neglected or entirely ignored. Fourth, a contextual structural force determined by the State induces conflict. Discussing the large corpus of political ecology is a kind of book view, so this chapter moves forward and takes into account the field view by discussing the Adivasi resistance against the developmental plans and policies which are tuned with extractivism.

5.3 What Resources Meant for Hos

In predominantly Adivasi regions, most Adivasi resistance and protest movements in mineral-rich central eastern states – Jharkhand, Odisha, and Chhattisgarh – have been centred on their struggle to protect *jal, jungle, and jameen* (water, forests, and land). In a wider ecological perspective, these natural resources provide the physical, social, cultural, and economic bases on which Adivasis' lives, livelihoods, and cultural dignity depend (Corbridge, 2009; Jewitt, 2008; Karmakar, 2002). They are characterized not by this or that race, habitat, or religious practice but by 'the bonding fabric of kinship and joint ownership of the natural resources from which they make their living' (Ratnagar, 2003, p. 17). Moreover, predominantly Adivasi regions in India have been characterized by rich land and the poorest people due to overexploitation of the region's natural resources to benefit the centres of power and wealth elsewhere (Baviskar, 1997b, 2003; Padel et al., 2013). In Jharkhand, natural resources such as forests, rivers, and mountains existed till recently without too much pressure to financialize them. The Ho Adivasis have treated them as common property resources for generations. The question of frontier expansion starts from the Hos' arrival and colonization of the territory, giving them a sense of identity (Sen, 2011), also known as a Hodesum (Dalton, 1872), as mentioned in the Chapter 4: How did the Hos arrive in Singhbhum and settle the land after clearing the forest? They had an idea of the common land and forest land. Their settlement of the land and claim of the Khunt Kattidars (original settlers) show that they were the recognized pioneer frontier of the region. Their claims were not contested in earlier days, but after the formation of the British Kolhan Estate in 1837, the institutional pattern of resource appropriation started in this region. The colonial State was the first to restrict tribal rights, control, and access to natural resources in 1885. Through the Indian Forest Act of 1878, and 1927, the colonial State exercised its power to generate more revenue and utilize the natural resources, mainly forest and land for resource requirements of the military, commercial, and industrial sectors (Kapoor, 2012). The Ho Adivasi was transformed into a tenant (Sen, 2011) that generated a concept of fixed private property.

Earlier, they had the idea of communal property in the form of common property delimited with their socialist value, which cherished the concept of ownership by the families or by a group of agnatic families. The communities occupied the region and always meant to keep it for themselves. Each family made clearances in the virgin forests, which came to be called Hatu, later known as Khuntkatti Hatu (already discussed in the Chapter 4) or the family's village. The boundaries of the village were laid by the old Adivasi patriarchs. They regarded the village as sacred and inviolable and believed that boundary gods (*Siman Bougaka*) kept a vigilant watch. Villages were demarcated up at four corners of a selected tract, and straight lines were drawn across the tract from one point to the next. These lines were connected through four bonfires, forming new boundary lines of the village. Within the territory of that village, all the land, cultivable and arable, waste, hills, jungle, and streams – everything above and underground became the common property of the village family. One bit of jungle was specially reserved for the village gods (*hatubongaka*), also known as Sarna. Verardo (2003), through her ethnographic exploration, shows how the 'nature-man-spirit complex' among the Ho–Mundas of Kolhan–Porahat works out on the ground. She states,

> They were spiritually connected to the land they settled. The process started during forest-clearing activities. Before the cutting of tree, the pioneers approach the so-called "spirits of the forest" (*buru-bongako*). ... The first person who cut the first tree would have to consult the bongas. If spirits are not consulted, or proper worship is not done, then misfortunes will happen.
>
> (Verardo, 2003, p. 89)

Worshipping forest spirits and ancestors into the root of the territory can be termed as the 'practices of rules' and the 'pact of the past' (*sidereyaniyam*). Pioneers stipulate a 'pact' (*niyam*) with forest spirits, by which protection is begged in exchange of the 'sacrifice' (Ibid: 89). Reciprocal duties and obligations to clan spirits are established through ritual sacrifices, and a close identity between territory, people, and clan-spirit are cemented. The pioneers become headmen and specialists in sacred practices, and these offices remain hereditary. Mundas' position (original settler family of the villagers) has been aptly described as chief among equals. They also obtained assistance from other villagers (brothers). Dalton (1872), in his book Descriptive Ethnology of Bengal, argued that this region had no Raja (King) during their takeover of the region known as Chutia Nagpur, which was a combination of small, allied states. Munda (literally means Head) also used to be the chief of each village (as a village often consisted of one killi, the inhabitants were all of Munda dignity, and hence it became a name for the whole tribe).

Their village-sasan with the rude stone slabs (*sasan-diriko*) that guard the mortal remains of the ancestors (*ha ram-ho roko*) of the village family is even to this day a favorite meeting ground of the once almighty Panchand on occasions of public importance. This is a kind of sacral value. It is conceptualized as 'any value that a moral community with hints or clear depiction treats as having unlimited or transcendental importance that precludes comparisons, tradeoff, or any other blending with bounded or secular values' (Tetlock, Kristel, & Beth Elson, 2000). These values are human life, health, nature, love, justice, dignity etc., and they are absolute and inviolable, and trading them with other monetary value is considered taboo.

Sasandiri (stone slab a symbol of self-governance) and *Jahira* are the sacral spaces of the Hos. These are a vital part of their ecosystem and landscape, and these are the cultural representative of ancient human culture. Their ancestral graveyards (*sasan*) stand as icons of the sacred relationship between pioneers of the village and its territory. *Ham hokonentaredum tanako* (ancestors have been sleeping here) – it claims territoriality among them when they argue this land belongs to us. It can be termed as 'customary title deeds'. In short, 'both the development of clan graveyards and adherence to ancestral spiritual practices legitimize Mundas' territorial claims' (Ibid.: 96). Jahira (forest) cutting of Sal trees is a vital part of their traditional conservation practices. The Adivasis' socially devised restraint displayed in their management of forests reflects the nature of these social formations (Corbridge, 1988). Sarat Chandra Roy reported their traditional way of using forest resources:

> It is in the months of Chait or Baisik (March to May) before the rains set in, that in many villages the Munda and Pahan, on a day appointed beforehand, lead the villagers into the village-jungles and the necessary fuel and timber for the year is cut down by the villagers from a specified part of the jungle, leaving the other part or parts to be similarly dealt with by rotation in successive years. And the wood thus cut down is then taken home by the villagers according to their respective needs. By this prudent use procedure, the village jungles can never be devastated.
>
> (Roy, 1970, p. 63)

The local ecology shapes Hos collective imaginations, belief systems, and institutions. The socio-economic and political organizations of the Hos society are determined by descent, kinship, and land tenure. It is embedded in a large number of *killi* categories (Yorke, 1976). The local lineage is initially established by the original ancestor who founded the village. On the whole, clan (*killi*) interrelationship constitutes the prerequisite and basis of all social and ritual interactions. There are villages with single and multiple *killis* spread into various Hatu (Ho village) and tolas. All these villages have had a specific kind of production determined through culture and have a homogeneous influence on all peoples. Their lineage mode of production is mainly structured and shaped through the customs at the village or community level. These types of production systems are the first stage of class societies. This mode of production is varied and numerous (Amin, 1974). The features of this mode of production are the absence of the commodity exchange and, correlatively, the distribution of the product within the community according to norms and rules related to kinship organization. Their pre-capitalist social formations have an ecological culture that must be seen as a part of social relations and forces of production. The popular assumption about Ho culture was that this society had been homogeneous and able to advance customs, and their economy depended on agriculture, forest products, a barter system, and frugality of wants (DasGupta, 2011; Singh, 1978). They are much more ethnically cohesive and politically empowered through their sacral polity and have had a strong attachment to land, forest, and water bodies.

This community developed a specific mode of production system which had made Singhbhum (Kolhan) 'a secure asylum' for them much before the first British colonial contact with the region during the 1817s (Dalton, 1872, p. 164). The British defeated and subjugated the Ho in 1836 after the great Kol rebellion in 1831–1832. As discussed in Chapter 4, their economy was self-sustaining and self-containing (Sen, 2011). The commodification, territorialization, and politics of conservation, along with land acquisition and its resulting resistance, have a long history in the district of West Singhbhum. It has been continuing with the process of State formation. As State formations rub up in Adivasi mineral-rich areas, their encounter with the State becomes vast and extended beyond our discussion, but it can be portrayed through their day-to-day encounter with displacement, land acquisition, mining lease extension, forest officials, and alike. However, an incident of land acquisition, commodification, and cases of undertrial of those who opposed and resisted the developmental projects is prevalent in Jharkhand. Hos have experienced the tyranny of development in the district of West Singhbhum from colonial times. It resulted in many uprisings and resistance. Some of the significant resistance in colonial times was Kol rebellion and Birsa movement, which was followed by various major resistance movements like Jungle Andolan of 1978, Swarnrekha Dam movement, Icha-Kharkai Dam movement, etc. There has been a long history of the development project induced tyranny that continues till date. Hundreds of short-term and episodic resistance against every mining project in the region are still going on with varied tactics. It is because these people have been proud and very rebellious, and their relations with the exploiters' forces have never been harmonious.

The history of Ho Adivasis in this region has been a chronology of resistance. Jewitt (2008) has termed it an 'ethnic conflict or environmental conflict'. Their opposition and continuous resistance against the State-sponsored natural resource exploitation, forceful transformation, and economic development had created the 'social pressures' that created a politically meaningful Jharkhand region (Stuligross, 2008, p. 133).

Even though the situation has not changed much in this region, the ways of interaction, negotiation, tactics, and politics have changed. After the formation of the separate State of Jharkhand, local political parties lost their issues and visions that diluted their philosophical concept of resistance in

West Singhbhum, which was the primus locus of the Jharkhand Movement. 'Jharkhand' meant that Jharkhand is for Jharkhandi (original settlers). Jharkhand, an already-weak and ineffective State, experienced political instability until the establishment of Raghuvardas' government. These local political parties forget their connections with the Adivasi ecology and history and are easily co-opted by the national political parties. They have also changed local politics, the depth of the resistance, and their ability to negotiate or bring change.

5.4 State, Mining Booms, Illegalities, and Resistance

Adivasi resistance against the State from the earliest period transformed from episodic and violent forms to everyday resistance. Capital logic is contrary to human development's logic (Lebowitz, 2002, 2009). The logic of capital requires the 'sacrifice of the human end-in-itself to an entirely external end but the question is, why do the Adivasis sacrifice their land and resources? Is it for the growth of capitalism because their region is resource rich? Lack of investment to improve rural agrarian infrastructure and alternative livelihood options remain at the heart of the Adivasis' genuine grievances in these regions (Harriss-White & Prakash, 2013; Shah, 2010; Stuligross, 2008).

These processes have reduced these regions into 'production of marginality' to resemble that of virtual colonies of extractive industry where governance and the delivery of basic public facilities, one of the main fruits of State legitimacy, have nearly been absent, whilst super-exploitation of its resources have been at its peak for the rapid development of urban centres (Kumar & Puthumattathil, 2018). In this way,

> 'the supply areas excoriate to an increasing stagnation and underdevelopment. The people who are also a host community for the mining companies here continue to remain as "marginal" "disorganised" and "uninformed population", who make demands only "in traditional supplications, petitions, and complaints"'
>
> (Blaikie & Brookfield, 1987)

The State's logic of human development and development as freedom (Sen, 1999) or dignity has been subordinated to the logic of the processes of primary accumulation. In West Singbhum, DI through extractive industry is synonymous with displacement, fear, marginalization, and pauperization. Corbridge (1993b) has examined the development intervention in West Singhbhum and argues that nothing changed in this region of Jharkhand, only power was transferred from a colonial administration to post-colonial administration. The process of primitive accumulation to 'dispossession without development' (Levien, 2018) continues in this region. The cases are varied but dispossession has always been true since colonial times.

Habib (1995) mentions that a new kind of independent circulation of accumulation originating from the non-capitalist economy to the capitalist economy has started in colonial India. The Tata establishment, ACC cement factory, and many others from the colonial times are the best examples to understand the circuit of accumulation in this region. Far from being relegated only to the colonial past, however, we continue to observe a process of primary accumulation to 'accumulation by dispossession' or regime of dispossession. The processes of primitive accumulation also exhibit variability over historical time. In many cases, these processes of the accumulative developmental project are tied up with various forms of repression, coercion, and appropriations, which can be observed in mining, dams, and infrastructure. Conflict, resistance, and collective uprising are now observed in every extractive project (minor to major) that comes to this region.

Earlier the doctrine of 'isolationism',[2] which includes the acts and provision of the Chotanagpur Tenancy Act (1908),[3] the Fifth Schedule Area,[4] and the PESA were against the discourses and practices of full-fledged fiscal developmentalism in this area. The government and its market-oriented development projects and policies suffered from these resistance movements, which restricted the State from establishing a complete economic hegemony in the name of national interest and growth. Corbridge (1993b) observes this ploy and tactics of the State and mentions that the modern State follows the 'ideology of benign integration'.[5] The State does integration in the name of development for the management of dispossession and serves the interest of capitalist class, an art similar to governmentality (Li, 2010). The State has facilitated development at the cost of commodifying the term 'improvement', which can be understood through the work of Polanyi. Following Polanyi's (1944) words:

> The poor man shall be satisfied in his end: habitation, and the gentleman not hindered in his desire: 'improvement.' This formula appears to take for granted the essence of purely economic progress, which is to achieve improvement at the price of social dislocation. But it also hints at the tragic necessity by which the rich man's desire for the public improvement which profit him privately.
>
> (p. 34)

All successive elected governments have treated the Adivasis of this region in a contradictory manner. Jharkhand's advantage is that it has abundant natural resources and ecological diversity. The State follows the neoliberal growth model and wants specialization in mineral and ore exports to get more revenue and surplus. After being bifurcated from Bihar, Jharkhand State (homeland for tribals) immediately chose the Jharkhand Industrial Policy (2001), which continues with the State's appeal to the investor to invest in Jharkhand through the various official platforms. The State has also given a platform via

Table 5.1 Location of Mines in West Singhbhum District

Location	*Number of Mines*
Noamundi	42
Manoharpur	6
Jhikpani	16
Jagganthapur	8
Chaibasasadar	8
Tonto	1
Haatgamharia	3
Kumardungi	11
Aabru	1
Sonua	1
Majhgaon	4

Source: District Mineral Foundation, West Singhbhum 2017

'Jharkhand Momentum' to attract investors for development through mining. Beyond regulation and the State politics of promoting extractivism, the policy is often criticized at various levels. It does not address the local people's aspiration of 'self-governance' or 'swaraj'. The State has been focusing too much on expanding the mining sectors with a lack of vision for promoting private players with no real direction for local people and local entrepreneurs. The World Bank reacted to JIP 2001 by calling it 'unrealistic'. The Bank argued that it focused too much on larger projects. The Jharkhand State has become an active investment facilitator in the mining sectors (see Table 5.1). Only in a few years, from 2000 to 2003, the number of mining leases increased by 150 per cent (Mazaheri, 2016). The government has signed a flurry of memorandums followed by Jharkhand Momentum with 209 memoranda of understanding (MOUs).[6] From the very early 1900s, the establishment of TISCO followed by hundreds of mining projects till now came to Jharkhand State because of the region's potential of resource richness. Table 5.1 shows the number of mines in this district.

Natural resource extractive projects come with a 'gift of bundle' (Li, 2014), including jobs, money, materials, resources, and knowledge to enter remote areas of the Adivasi territory and leave just as quickly once the resource is exhausted. Political economist Harold Innis terms this 'cyclonics' (Deibert, 1999).[7] A mining boom for resource extraction was observed after the formation of Jharkhand. Mazaheri (2016) reported that the mining lease area was the highest in any State of India in 2008. In the field area of West Singbhum, out of 18 blocks, six blocks and their 31 panchayats have suffered from extractive capital and State tyranny (see Table 5.2). Except for these panchayats, there are many minor extractive projects which the mining office of West Singhbhum does not report. Table 5.2 shows the mining-affected blocks and panchayats, which shows how the presence of the State and

Table 5.2 Mining-Affected Blocks and Panchayats in West Singhbhum District

Block Name	*Area Affected (%)*	*Total Number of Affected Panchayats*
Noamundi	67	12
Manoharpur	58	11
Jagganathpur	6	1
Jhikpani	43	3
Chaibasa	14	2
Manjhari	20	2

Source: District mines department, West Singhbhum

extractive capitalists is also experienced in this region through these projects. The experience can be varied from time to time and occasion to occasion. The State's activities can be episodically felt vibrantly at the time of the public hearing, lease area extension, lease renew, investigation of field sites for afforestation or deforestation, environmental clearance, and election.

Only after 2016, in West Singhbhum, the acting commissioner of Kolhan directed a formal closure of stone crushers, bricks kin, and mining activities in Dalma, an eco-sensitive zone of the Kolhan region. The State once permitted mining companies to extract the resources rollback its decision after the alarming situation of the region. Kolhan commissioner directed district mineral officer to coordinate the department of environmental, Government of India. It shows how there is a gap between the two different sections of bureaucracy. Earlier, they formed a task force to close the mining, but mining continues.[8] This event needs critical elaboration. People of this region and nearby region often complained about the illegal mining, ecological degradation, and pollution due to crushers and mining transportation, but their voices were not heard. Dalma Wildlife Sanctuary was declared an eco-sensitive zone after the alarming deterioration of the ecosystem of the region by the Union Ministry of Forest and Environment. It was notified to keep out of bounds the whole forest of 522.98 square kilometres, including 85 villages in the forest and 51 villages in its periphery, for any permanent structure(s). This replaces the earlier notification, which had considered the area up to 5 kilometres from the boundary of the protected Dalma Sanctuary in the eco-sensitive zone.[9] The eco-sensitive zone covers an area of 377.89 square kilometres that falls in East Singhbhum and another 145.09 square kilometres in Seraikela-Kharsawan District. The State once allowed mining in this region but after the ecological degradation it converted this region into a conservation zone by reclassifying and remapping the territory.

Government agencies are continually reclassifying and remapping territory to account for how people and resources are governed. This kind of 'greenmentality' is an oxymoron. Once mapped and reterritorialized, villagers are forced to leave their village. Both extractive politics and conservation politics of State produced this kind of marginality. In other forest regions of the district, Porahat and Saranda, mining operations are still going on that disturb the ecological equilibrium of the region. Mining-induced conflict between the Adivasis and State and disturbed the relationship between the Adivasis and the wildlife. Mining-induced deforestation has changed the route of elephants and other animals. The Forest department of this region proposed a four-elephant corridor to lessen the conflict between elephant and human.[10] Future plan of Forest department of this region reported that mining has caused huge breaks in the continuity of the habitat. It has spoiled the cover area of the forest which affected the livelihood of the local people. The forest, the religious and sacral landscape of the Hos, became a site of corporate encroachers. It is reported that mining companies have seized huge profit by encroaching on unallocated forest land. Illegal mining was so common in this region. The commission of Justice Shah also reported it. The Government of India formed this commission to enquire about the mining status, illegal trading transportation, and the impact of mining. The commission noticed that huge deforestation, illegal mining, and unlawful trading of iron ore and manganese are very rampant and mining operation is not transparent (Shah, 2013). Illegal mining has been rampant among legal lease holders, yet no significant action has been taken to address this issue. The central government has told the court that extending the tenure of the Justice Shah is unwarranted. Most surprisingly, the central government did not submit the final report to Supreme Court by mentioning that placing the reports before the court would 'impinge' upon the jurisdiction of Parliament and pre-empt the statutory provisions.[11] This commission also pointed out that 18 mining companies were running under deemed extension without environmental clearance, and 22 companies were carrying out mining in violation of laws. Out of 42 mining leases of iron and manganese ores, encroachment of various mining activities was observed in the case of 25 mining leases. They also reported the massive indifference of 53.41 MT of iron ore production in data of State government and Indian bureau mines, which shows how mining was rampant by ignoring the environmental concerns in Jharkhand.[12] Most of the mining companies were operating in West Singhbhum because the region has a vast amount of world-class quality of iron and manganese deposits. It is reported that iron ores worth over INR 22,000 crore and manganese ore valued around INR 138 crore were illegally extracted by the big mining companies, like the State-run mining companies SAIL, and private companies like Tata Steel and others.[13] Table 5.3 shows the illegitimacies of some of the corporate companies which promised to improve the region.

It has been alleged that the above-mentioned mining firms in the Saranda region of West Singhbhum have produced and dispatched excess quantity of iron ore against the permissible amount under environmental and mining laws.

In this vein, Mukesh (2012b), in his reporting in the English national daily *The Telegraph* mentioned the letter of the registrar, Sudhir S. Shah, who questioned the role of the concerned authority for not taking any judicious action

Table 5.3 Mining Players and Their Illegitimacies

Name of the Company	*Mines*	*2008–2009 (in MTpy) Approved*	*2008–2009 Produced*	*2009–2010 (in Mtpy) Approved*	*2009–2010 Produce*	*2010–2011 (in Mtpy) Approved*	*2010–2011 Produce*
Usha Martin Ltd	Bijaya-11	0.65	1.833	0.65	2.324	2.4	2.572
Shah Brothers	Karampada	0.612	0.715	0.09	0.679	0.9	0.972
P. K. Jain	Thakurani	1	2	1.3	1.91	1.3	1.68
OM & M Ltd	Ghat kuri	0	0.18	0.72	1.2	0.72	0.943

Source: Mukesh, 2012a

against the mining illegalities done in the years 2008, 2009, and 2010. Although former rural minister of the Government of India, Jairam Ramesh, was against the mining done by private players, he himself, as a minister of environment and forest, gave environmental clearance to Usha Martin Ltd, a private company, to operate mining activities. However, his point of view also changed with the passage of time. He might have realized the destruction the mining company had done in the name of development. His successor Jayanti Natrajan again neglected his suggestion of not giving clearance to private companies. Many companies, including Jindal, Rungta, JSE Steels, etc., were given clearance for mining (Dungdung, 2015). The environmental clearance given to the mining companies shows how economic growth and maximization of surplus and profit have been more important for the State and capitalist players than the lives of the local tribal population.

This maximization and diversion of the forest are still going on in the name of development and betterment of the region. It is a fact that the State and capitalist players always attempt to convert the surplus value of accumulated land and resources into profitable value, whereas the Adivasis having a sacral subsistence polity and imagination always try to resist them (Naik & Tiwari, 2007). Two societal formations interact: The State always mediates the dominant production system and societal formation and shifts its responsibility to the corporate to take care of the community. The State in return favours corporate through changes and modification, even manipulation of laws and acts. The State government and its apparatus ignored the environment and other obligatory provisions at the time of allocating and giving recommendations to the mining companies. During a personal conversation with Saryu Roy, an environmental activist and a senior minister of the Raghuvar Das-led Bhartiya Janta Party government, he conveyed that the previous administration assigned land without conducting an initial survey of mineral availability. This resulted in the transformation of forested areas into de facto mining zones.

Major acts such as the Mines and Minerals (Regulation and Development) Act of 1957 (MMDR) and Mines Concession Rules of 1960 are the basis for granting prospecting licences and mining leases for all minerals. This is done in accordance with other laws like Forest and Conservation Act, 1980. One of the State illegalities was observed when the State government did not do an initial survey and recommended the applications of 34 companies for mining. Section 5.2 of MMDR Act has a provision that a State government before recommending will have to conduct an initial survey of the region about the availability and stock of the minerals. Even then out of 34 companies, three mining players got land for mining after getting permission from the State government of Jharkhand. These companies were Balaji Industrial Product Limited, Abhijeet Infrastructure Limited, and AMM Power and Limited Company. The areas and names of the regions were, respectively, followed: 420.95 hectares for Balaji Ltd in Bokna region, 138.50 hectares for Abhijeet Ltd in Katoria region, and 383.54 hectares for AMM Steel and Power Ltd in Silpuji region. Out of three, except Abhijeet Ltd, the central government of India also gave prior permission to two capitalist players for mining, but when they started to mine in the allocated region, it was found that stocks of the minerals were not available.[14]

This is an ideal case that illustrates how the regulatory system disregards its legal provisions, leading to forest degradation, deforestation, and the displacement of forest communities, ultimately depriving local residents of their forest rights. This act was amended in 2016, showing a neoliberal State's attempt to empower and allow mining companies to dump their waste outside the lease areas.[15] This amendment addresses the ambiguity that came after the decision of the Supreme Court in the case of *Samaj Parivartana Samudaya & Ors v State of Karnataka & Ors* (a Karnataka case) and subsequently in the case of *Goa Foundation v Union of India & Ors* (a Goa Foundation case). It empowered mining companies to fully exploit the whole lease area while dumping their waste on the local people, but the local and national political parties did not oppose it. This act has a history of the amendment. It was earlier amended in 1994 and 1999 to accelerate the flow of private capital (domestic and foreign). George (2014) mentioned how Jharkhand was in a hurry to attract more companies to mine. She reported:

> The State government of Jharkhand would encourage joint venture projects with the State Mineral Development Corporation (SMDC), especially in mining. The private sector would be encouraged to take up mining activities in the state. Mining lease applications along with project report and all relevant documents would preferably be disposed of within 60 days of filing such applications
>
> (George, 2014, p. 33)

Brinda Karat, a noted communist leader (2012), mentions a report compiled by Ernst and Young, according to which 4.9 lakh hectares of land were given out in mining leases in 23 States by the end of 2009 and 95 per cent of the leases comprising 70 per cent of the land were given to private companies.[16] She warns that liberalization of the mining and mineral sector will ruin the Adivasis' constitutional and customary rights.

5.5 Adivasi Resistance and Politics of the State and the Corporate

The struggle of tribals is a part of the democratic struggle against the capitalist path of development; the democratic movement in the country has to recognize in the tribal people's struggle a powerful ally – 'an ally with tremendous revolutionary potentialities for the realization of national democracy and socialism' (Bardhan, 1973, p. 57).

Bardhan's statement is relevant to understand the tribal struggle against the modern State. They are the sufferer of the making of the contemporary State, and when they resist, they frame their historical experiences of marginalization and injustices (Cavanagh & Benjaminsen, 2015, p. 730). Whenever any mining project was initiated in West Singhbhum right from colonial time to now, it had started with the surveys of geologists and engineers who conducted the initial surveys for the State and companies. There are many similar cases of land accumulation and resistance.

One of the collected cases is from the Nakahasa village. Here, the State regulatory system is only experienced through mukhiya and some populist schemes like Public Distribution Schemes. There are no branches of ATMs, banks, or post offices in the village. This region is predominantly forested with agricultural land without the State's services. There are no good roads, and the long-standing electricity polls only confirm that the State agencies are present just for namesake.

One day a limestone company got the lease in the remote areas of the Nakahasa, the countryside, and sacral polity of Ho, which has fostered strong community ties with ecological and cultural conservation, started to resist the projects, but, initially, they failed. The land was acquired and operationalized with the help of the bureaucrats of the government mining department and block level. It was done without the prior permission of Gram Sabha. As limestone mining started, four high-capacity diesel generators were used to pump the water to extract limestone. Soon the villagers of Nakahasa and nearby villages Bambebesa, Mauda, and Banaikela realized that the groundwater was depleting. Not only did the villagers face a shortage of water, but their agricultural land was also filled with dumped waste stone that hampered their agricultural activities. People of the Nakahasa came seeking help from the Gram Sabha, and, under the presidentship of the local Munda, mining operations discontinued. They submitted a copy of their decision to the deputy collector and other local government officials.

During a conversation with the villagers, they said that allowing extracting limestone from our village again would turn the rest of the land barren and arid. It would destroy our sacred grove (*sasandiri*), village harmony, and agricultural lands. As resistance continued, mining company recruited some agents and paid 'upper' to manage the villagers to ensure the least resistance to start the mining operation by bribing a few village 'uppers' and local people secretly and paying some compensation to the four villagers whose land were chosen to mine along with government land. They did it because they also knew that the actual extraction of limestone was not possible without the villagers' consent. After giving compensation, they approached and managed to take the positive support of the Munda to establish the mining company. The management of upper and local leaders was to break the solidarity of the Adivasi protest, old capitalist ploys and tactic that can be conceptualized as 'hold strategies'[17] (Brock & Dunlap, 2018). Hold strategies failed when villagers reunited after knowing the tactics of the industrialist. They opposed and even beat the agents of the local industrialist. As Devi (1981) observed and briefly stated, 'To keep the pattern of exploitation running smooth, tribal solidarity must be broken into atoms, but it did not succeed in this village' (p. 1597).

Days went by quickly; one day in August 2013, government officials, police, and industry agents again came with mining equipment, tools, wires, and a generator. The State presence is only experienced vibrantly when State officials come to negotiate or mediate between corporate and villagers to take their 'consent' to start the mining operation. As they entered the middle of the village, the information spread like wildfire. About 50 villagers gathered at the mining site and started to protest the move of local State officials and agents. A young Ho lady, who was a Para teacher, asked the agents and officials, 'Why did you people come, and what do you want? We have already taken a decision through village assembly (Gram sabha) that we do not need mining and your accountability for the development of this region. We will not allow you to mine anymore. When there were heated arguments, State officials started to favour the industrialists. Government officials asked the agitators to be patient and declared that these lands were already leased out, and adequate compensations were given to four people who had given their land. The officials asked the villagers why people were resisting. The villagers shouted collectively, and one of them mentioned that the land belonged to the village, not to an individual. If mining started, it would degrade the village's life choice, ecosystem, and agricultural fields. Earlier mining in this region and the nearby village was a curse for them. The whole village had already faced the problem of water scarcity and pollution because of mining dust. It had already destroyed some of their agricultural fields and transformed the land into Death Valley of water. They would not allow it to happen again. Negotiations between the officials and agents with the villagers failed, and villagers seized the equipment and tool of the company and forced them to leave the boundary of the village (Jyoti Bari; personal interview 11 August 2016).

Why do people resist? A Mundari song collected by Devi (1981) from the field of Singhbhum explains it well. It states, 'Our land vanished like dust before the storm, our fields, our homes, all disappeared. The ones who came were not human beings. ... The forest disappears; they make the four corners unclean. Oh, we had our ancestors' graves' (a Mundari song recorded by Mahasweta Devi).[18]

People who resist know that mining means losing land and forest; their experience dates back to the colonial times. The notion of tribe in itself 'is being transformed into a tribe for itself. Not only are the tribes busy creating a self-image, consolidating their ethos, culture, and philosophy but they have also started their movements and founded their own organizations to combat injustice' (Srivastava, 2008). It is a prophetic activity. 'The resistance of the affected communities points to the fragility and inequality of the democratic process and also highlights the capability for self-defense and self-organization in these communities' (Poma & Gravante, 2015, p. 59). The aim of such resistance movement is to keep the mining activity out of their village. The above-discussed case of Nakahasa is an ideal example of grassroots resistance against the State and mining companies.

This case highlights several themes: First, it illustrates that it is the right of the villagers to decide what should be done in their village, and that the district mining office, State bureaucrats, and leaseholder company should not decide the fate of the local people through extractive projects. Second, it highlights the alternate imagination of development and their countervailing power. Third, it provides an example of how today's Adivasi (Indigeneity) resistance is a strong force to counter neoliberalism. Their resistance to mining is based on their custom, according to which the land belongs to *killi* and village. Many scholarly works (Denny, 2008; Verardo, 2003; Yorke, 1976) have discussed the ancestral and spirit world of the Hos in great details. Every Ho household has a sanctuary called *adding* where ancestral spirits dwell, and a 'Ho village' remains home to ancestors within its boundaries.

The relation between the living and the dead reveals that ancestors are integrally involved in the social and ritual lives of a Ho, which is also integral to their sacral polity.[19] Hos imagine themselves as living within 'a religiously defined boundary that is realized through a contact with the village guardian spirit', or hatubonga, who resides in every sarana (sacred grove, also called desauli of a Ho/Munda village). There are other burubongako (spirits residing in wild forests and hills) of whom Hos feel threatened (Yorke, 1976, p. 55). Customs of Ho prohibit the transfer of land. They always prefer their customs and ecology to the monetary economy that points to the idea of community solidarity above the individual profit, but it does not happen all the time. The State has forced them to adopt a market-based ethic. Several schemes of the State and the corporate social responsibility (CSR), an industry policy, are working in this direction. CSR policy is actually colonizing and, therefore, hiding (colonizing) the social reality of the impacts of mining. Free health checkup, involvement of cultural programme, recognition of the cultural festivals of local community, organizing football and other sport for Adivasis, establishing private school, and building of infrastructure are some examples of CSR. It is a strategy that is being used to win the heart of the people around the mining projects to lessen the resistance. Sometimes they win, especially after the formation of a new State. The State facilitation and capitalist penetration continuously shape the societal vision of the local community. State and capitalists jointly adopt hard and soft techniques (Brock & Dunlap, 2018). The soft and hard technique can be observed through the corporate-political entanglements at the local level that also work at the time of lease allocation and lease operation to lease renews. This nexus and merging of State and corporate apparatus signals towards the integrated cooperation that symbolically addresses the resistance. Soft politics are tied with CSR, a strategy and information operation activity to build a bond and affection with local people. It has been used for a long time to change people's attitude.

Resistance against the mining also shows the environmental ethics of the Ho community that challenges the accumulative projects of the State, taken together, cases observed from field revealed that Ho community ultimately proves the key to resistance because it is not reducible to the logic of capitalism. It also shows that if the community resists and does not get atomized, they can challenge the neoliberal development model. No project will succeed if the community refuses to bear the cost of production. Mining and extractive industry talks about prosperity, job, and poverty, but those who challenge them know that such prosperity is only the short-term boom of an export sector. Unlike forest and agricultural production, the life of the mining sector is limited. If once biodiversity and local ecology are devastated, then getting an unskilled job, monetary gifts, and infrastructure cannot compensate their loss.

There are many similar official cases of land acquisition that show how the State's involvement to facilitate capitalist player is a fraudulent hegemony. In West Singhbhum, there are ample examples of such cases dealing with land grabbing and acquisition. Kishwar (1987) reported how the mining operation was going on in West Singhbhum:

> A quarter of mining activity in India is carried out in Singhbhum [Kolhan]. Formerly mostly forested land today has been brought under cultivation or reduced to rocky waste by increasing deforestation... Most of the land remains with the Hos due to lack of irrigation, electricity, and low productivity in agriculture. ... Besides large scale land alienation due to mining, industrialization and big dams, external political structures, owing their allegiance to exploitative, hierarchical government machinery, the Hos are

> forced to practice non-viable subsistence agriculture, with soil fertility destroyed, no irrigation and other technological inputs. ... Most families supplement farm income with other sources – working other people's land, collection and sale of forest produce, and seasonal migration as casual labour in mines, construction industries, brick kilns and other worksites. Children perform about 90 per cent of the total labour in villages. ... Most lucrative jobs in the industries have gone to non-tribals. ... All the families grow one major rice crop a year which they sow during the monsoon. However, when rains fail or are delayed, there is not even one crop a year. Even one year's monsoon failure causes scarcity. Often, the crops fail for several years in succession. As a result, food scarcity is a chronic problem for most families.
>
> (Kishwar, 1987, pp. 95–96)

Here it is pertinent to discuss the politics of scarcity. In West Singhbhum, scarcity is not due to climate changes, drought, and less monsoon. Earlier, if it happened, it would be a social scarcity, and they would have had an alternative regarding forest products. However, here, what is most alarming is that the scarcity is arising from economic crisis caused by capitalistically produced scarcity of nature or production conditions. This is embedded with capitalist production relations that show that scarcity is not due to only bad harvest, but it is due to the capitalistic mode of operation that treated nature as a commodity and transformed our sacred river into polluted river and exploited land, forest, and water resources. However, the State translated scarcity as poverty, backwardness, and argue again for intervention through different capitalist projects to bring economic changes in the lives of the Adivasis. State plans and policies to bring economic change don't address the 'culturally formed wants' (Hefner, 1990). As Hefner mentions, 'economic change is never just a matter of technological diffusion, market rationalization, or capitalist penetration'. Deep down, it is also a matter of community, morality, and power (Hefner, 1990, p. 2). The question of power is deeply embedded with questions and implementations of corporate development projects. These corporate-driven development projects show tendencies of 'casino capitalism'[20](Sinn, 2010).

If a settlement is not a 'host community' today, it may nevertheless turn out to be the custodian of some hitherto hidden resource desired by businesses tomorrow. And if some indigent people are not construed as stakeholders immediately, they may yet aspire to this happy condition at some point in the future, when the wheel of fortune has turned in their favour. The odds of this happening are probably, in the cold light of business reality, somewhat slim, but the 'beauty of the convoluted CSR apparatus is that it continually holds out the possibility that the poor may – one day – get lucky' (Sharp, 2006, p. 221).

How can a community be lucky if its ecological resources and landscape are degraded through mining? It is translated into an economic crisis that can be understood as a second contradiction of capitalism (Connor, 1988). State and corporate employ the discourse of sustainability and CSR policies to address the crisis, which often leads to resistance.

Even the plans that State and corporate companies make through CSR to incorporate the local Adivasi social system into the dominant economic system are lacking in terms of participation and procedural justice. They fail to satisfy the need and cultural wants that are indicated in the slogan '*jaan denge lekin jameen nahi denge*'[21],chanting at the time of their reaction against the State and corporate nexus.

'Where there is resistance, there is power' (Abu-Lughod, 1990, p. 42). Power may be understood in a Foucauldian sense as 'something practiced in all social relations throughout the society, on all levels and in indeterminate struggles, negotiations and changing relations of forces' (Vinthagen & Johansson, 2013, p. 2). The following case, reported by Stuligross, showed how mining lease and operation is undertaken:

> The case of the Sarda Mine in Chaibasa is typical. The villagers were not willing to part with their lands, which comprised their sarna [sacred grove], agricultural lands and house sites. In order to obtain the lease, compensation and agreement papers had to be signed and produced before the mining department. These signatures were obtained by making some tenants drunk and dragging and beating others. Those who still did not yield had a taste of police lathis [nightsticks] and jail life before their thumb impression on blank paper could be bought for a nominal sum. … Practically, every mining lease in Singhbhum is purchased in a similar way.
>
> (Excerpt from a petition filed by the Singhbhum Labor Union Chaibasa during the early 1980s, quoted in (Stuligross, 2008))

In the context of struggle over resources between State and community, 'people may invest in meaning as well as in the mean of the production- the struggle over meaning are much a part of the process of resource allocation as are struggle over surplus or the labour process' (Berry, 1988, p. 66). These struggles are animated by local history, mediated by cultural idioms and political imagination that show how local inhabitants have pursued different practices in defense of their livelihoods. In the context of the developmental project, this resistance against the State's intention to develop the Adivasi region shows that State features are internally differentiated in nature. If the State is internally differentiated, it becomes a site of struggle (Mitchell, 1991). This 'everyday advocacy'[22] has very nominal effects on the State. The State is a sovereign theater in which 'resources; property rights and authority are struggle[*ing*] over' (Watts, 1989, p. 4). The agenda of the State apparatus is to pursue the dominant neoliberal agendas in this region. The opposition against any DI is advertised as an opposition between a monolithic State and an undifferentiated peasantry (Moore, 1993). The State can also be described as 'those aspects of the governing administrative and coercive apparatus that are experienced as external yet hegemonic' (Tsing, 1993, p. 26).

One of the cases that is relevant is that of Hat Gamharia block of West Singhbhum, in which more than a hundred people of Pauipi, Kochra, and Jamdiha came near Joraibai to oppose the area extension lease renewal of ACC Cement Company for mining of limestone.[23] In that general meeting, people from more than 16 villages of the affected panchayat joined the forum to oppose the extension. During an interview with the villagers, one elderly member of the village mentioned that an additional deputy collector allowed ACC Company to choose the land plot through an order. Once villagers got the news, they collectively decided to oppose it. The company's people illegally surveyed the agricultural crop land. A general assembly (meeting) of the villagers under the shade of a tree in Joraibai condemned this bureaucratic decision.

One important point came up during a conversation among the local people; when it was asked as to why they resist, elderly villager replied:

> State and not even company informed us about the survey. They came suddenly and started to demarcate the plot. They are a big giant and mined in nearby regions for many years. Still, those region is now suffering from water depletion, pollution, and degradation of the local resources and customs. Why we give our place to them for mining. We don't want the same in our village, so we are protesting.
>
> (From field notes)

An old local, Manki (Budhan Singh Sinku), of the region termed it a 'conspiracy of the government', a conspiracy that has an impact on thousands of local people. Munda of the Jamadiha (Sunil Sinku) mentioned that mining and factory establishment would bring outsiders (dikus). The villagers believe that this will displace them from their ancestral land and force them to beg or do manual labour. One of the members of Zila Panchayat, Sumitra Sinku, said that government officials allowed the company to choose their land plot without permission from the local '*raiyati*'people. She made an allegation that *sarkar* gave preferential treatment to the companies and added that if it continues at the same pace, it will alter their '*Adivasiyat*'.[24]

The field cases have shown that the the customary chiefs and elderly people of Ho community of these regions believed that government bureaucratic machinery often helps corporate to finalize the land deal while ignoring the mining and environmental laws. Additionally, they ignore Gram Sabha's decision and local political bodies. The State bureaucracy is caught in a strange combination of a Reichstaat and arbitrary power of slow legalistic and legislative procedure that favour capitalists. Before allocating area for mine, State officials generally declare that acquired land is not an agricultural field and mining will not harm the farming; but this is not a fact. The State's classification of land as marginal, waste, or degraded for accumulative projects has led to a 'very rough, sometimes misleading, representation of actual existing rights to land' (Scott, 1998a, p. 47). Thus, it is a false assumption that classified or categorized land is available and appropriate for accumulative projects (Schneider, 2011). He argues that, in reality, this land belongs to the community. By disregarding communal land use practices, the State dispossessed the local community by accumulating land for mining and resource exploitation.

These cases from the field also show how the State-led developmental activity has failed to justify the aspirations of the commons. The geological survey and mining department jointly traced huge stocks of iron ore deposits in a village, Kotgarh, without prior permission of Gram Sabha. It scared the villagers. A local elderly villager of Kotgarh said that the coming of the industry gave them anxiety, fear, and tension when they thought that diku (outsider/exploiter) would come one day and forcefully displace them, with the help of the State machinery (Police), in the name of development. No one would hear their voices. In their district, this is continuing from colonial times. Adivasi people had a lot of aspirations associated with the formation of the new Jharkhand State, but successive sarkars[25] betrayed them. He further said that they fought for absolute autonomy; and when they got Jharkhand, they thought that they had got independence from the exploitative domain of Bihar. They aspired to develop their community according to their means and necessity. 'But it does not happen because we do not get the real Swaraj [Independence]. Jharkhand is made and meant for us, but the Sarkari [Government] machinery often favours the capitalists and mining companies. Money is raining in these areas, but it is not for us. Dikus and the local leader do not worry about the local ecology. The whole process of Sarkari machinery treats us as strangers in our land' (Personal interview, 12 February 2016). The elderly man represents the fear and anxiety of the villagers. Fear, anxiety, and agony are very common to the entire region where these developmental projects are transformed with accumulative projects.

People know well by now that the company would give them nothing; they come to mine and then go back. For mining, they use different ploys and tactics, like promising to provide employment, infrastructure, and other facilities. It is an attempt to transform them into an agent of production who cannot dare to resist any more if they are co-opted. For co-optation of local community, State uses different discourses that can be conceptualized through Gramsci's concept of 'common sense' (p. 322).[26] It elucidates how natural resources will be best managed and used by the State that will benefit the local community. It only helps the State to achieve hegemony through its apparatus and developmental discourses. The State's process of public hearing before mining starts or renews a kind of drama to take Adivasi consent tactfully. Once cleared, the State transforms the tactful consent into the social life of the common people where the mining operation is proposed to start. The imposition to consent for dominant groups and State leads to assimilation and co-optation.

Co-optation is a process whereby a more substantial group like State and mining companies subsumes and assimilates the lives of the weaker community. It starts when the alliance of different actors like State and corporate can suggest compromises and force the weaker section to change its original discourse and demand. Faubion (2000), who has edited the work of Foucault's lectures on power, argues that the transformation of man into labour, as agents of production, was synthetic and political. It was brought through the discourses of development (Escobar, 2011). Faubion mentions that there is no hyper profit without infra power. Infra power is not a kind of political power, but a whole set of power of little institutions situated at the lowest level. Infra power, a prior condition of hyper profit, in establishing itself, in beginning to function, gave rise to a series of knowledge – a knowledge of the individual, of normalization, a corrective knowledge – 'that proliferated in these institutions of infra power, causing the so – called human sciences and man as an object of science, to power' (Faubion, 2000, pp. 86–87). State produced knowledge through developmental projects for co-optation of local community but those who are not being co-opted continuously resist against the State and companies. In fact, their resistance movements against the extractive capitals are also an attempt to negotiate with the State from the below. For resistance, they derive their strategies, tactics, and politics through the sacral polity and history that gives meaning to their activities, which can be appropriately called 'culture' (Geertz, 1977). Their resistance can be also conceptualized as land and territorial resistance. It means that the companies and State demand access to a means of production on their village land, and later treat the land as a territorial habitat (Sauer, 2012, p. 96). The State through several provisions of constitution guarantees the right and recognizes them as an original settler, but this society has experienced being a risk society. The feature of risk society is that governments and other power holders 'increasingly occupied with debating, managing, and preventing risk that has produced', which redefines society's relationship with potential threats (Beck, 1992, 2012). Social and environmental risks are apprehended in society as technical problems to be contained or controlled through, for instance, a management plan or an impact assessment before mining, transforming a forest region into mining zone. It is the State-hindering-alternate-ways of problem-solving that needs ethical solutions. Earlier, some regional political parties at the time of Jharkhandi movement had environmental ethics. They were worried about '*Jal*' '*jungle*', and '*Jameen*'. But it was an episodic event. Now in any instances of resistance or protest against mining, the involvement of the political parties is almost negligible. It was a fact that in all phases of

Jharkhandi movement, the contribution of local parties was very aspiring. These political parties and Adivasi leaders were fighting to save the *jal, jungle, and jameen*. It also gave them political benefit. Many of them became MLAs, MPs, and even CMs of the State but Adivasi agony was not heard by any of them. In this district, all the members of the legislative assembly belong to local political parties, and they all are Adivasis. After being elected as an MLA, they do not worry about local issues of livelihood, displacement, and injustice. Many of them belong to Jharkhand Mukti Morcha (JMM) Party, which has a history of engagement in abundant sets of episodic resistances during Jharkhandi movement. But as time passed, mainly after the formation of Jharkhand, their goals and objectives regarding mining and neoliberal development also changed and were co-opted by the neoliberal forces and national political parties.

Politics can betray, but customs cannot. Now the local Ho community has adopted other ways as they directly derive their power from the constitutional acts and State policies, which act as the safeguard for them. Their consciousness has risen due to some human rights-based associations and activist associations (*manch*) against the extractivism. Now the local resistance has shifted to the village level, and in many cases its philosophical foundation is based on liberation theology. Many associations like Jammen Bachao Samanvay Sameeti, Bhumi Bachao samanya Sameeti, Yuwa Jhumar, and Adivasi Krantikari Mahasabha are working on the philosophy of liberation. These associations and groups assist the conscientization and critical self-reflection of the Adivasis to understand the power asymmetry and developmental politics. It is a kind of awareness that is necessary for resistance.

A Chaibasa-based independent researcher associated with the Tribal Research and Training Centre (TRTC) argues, 'if there is injustice, there is resistance. In Adivasi resource rich areas, the struggle against the capitalist's forces and State may be understood as a democratic struggle' (personal Interview 12 March 2016). The work of these associations lays the groundwork for grassroots resistance by educating community members to become leaders, to see themselves as humans with full rights as citizens, and to organize the solidarity of the community against injustice. The activities of these organization are not only limited to one place of struggle or resistance but also engaged with other issues like land acquisition, mining areas renewal, and extension. They inform local villagers about the politics of public hearing. If a capitalist player is prominent and influential like ACC Cement Jhikpani, Rungta, and Tata and has a mine operation, then they raise only the issue of employment, pollution, ecological degradation, and other social benefits. They make their steady demand through regular *dharna*, negotiation, and rallies. It is a negotiation from below through everyday resistance in front of the company and the district collector's office. Their demands are made to the company administration through mediation by the State officials. The State is the most influential player, among all, to whom these

Adivasis submit their demands. John Munda, a leader of Adivasi Krantikari Morcha, has made an awareness campaign regarding the power of Gram Sabha in West Singhbhum. He often says. '*Na Rajyasabha, Na loksabha, Sabse bada Gramsabha*'. These organizations are working to empower the Gram Sabha through the direction of PESA and make people aware of their constitutional rights and different laws and rights regarding control and access to natural resources. It is an attempt to educate them regarding their power which will help them to act against capitalist encroachment. It is a kind of everyday resistance that concerns people accepting, abiding by, and questioning the norms and regulations of authority over allocating and accumulating resources in an indirect and unplanned resistance.

5.6 Conclusion

This chapter has enumerated how the State's rent-seeking behaviour is based on the neoliberal model of growth. Land for the tribal communities is intricately bound as a resource, livelihood sustainability, and place of their ancestor propitiation, social practices, and religious rituals. Over time their rights to land in almost all its forms have been eroded due to sale (often coerced/forced and illegal), acquisition by the government for public purposes and forestry measures, fraudulent transfers, forcible eviction, mortgages, lease, encroachments, and distress sales due to money lending. It lays the base of dispossession that resulted resistance. State's undemocratic practices and mining illegalities, as well as irregularities, have shown how State is a big theatre of neoliberal project where Adivasis are bound to bear the ecological cost of the production. Primitive accumulation, law manipulation, production of discourse, and employing the apparatus to favour the capitalist player is an everyday tyranny for the Adivasis. Chapter 6 will deal with the decentralization of resource governance and its politics at the village level.

Notes

1 In West Singhbhum, slogans like *Abua raj* (Self Rule in Village) were based on the foundation of Birsa Bhagwan. Birsa Munda is a class hero for them who fought against the colonial Raj. Ideological domination here refers to how the state imposed capitalist idea of living and development model and how it is being opposed by the defender of moral economy (Adivasis).

2 See Elwin (1943), His theory is known in anthropology as public park theory. It advocates that tribes should live in isolation. The doctrine of isolationism was very controversial. From 1940 to the debate of constitutional assembly, isolationism to assimilationism was the centre of all debates in the constitutional assembly. Elwin changed his earlier version of cultural segregation and emphasized a careful planning for tribal development (Elwin, 1965)

3 The Chotanagpur Tenancy (CNT) Act was enacted in 1908 after the Birsa Movement to govern land issues and prevent tribal land alienation. It is a kind of Magna Carta and blue print for tribals. It was prepared by John Hoffmann, a

missionary social worker. Its operation is effective in North Chotanagpur, South Chotanagpur and Palamau divisions, including areas under various municipalities and notified area committees. Section 46 of the CNT Act restricts the transfer of land belonging to Scheduled Tribes/Scheduled Castes and Backward Classes. However, a tribal may transfer his land through sale, exchange, gift, or will to a fellow Scheduled Tribe member and residents of his own police station area. Similarly, SCs and BCs can transfer land to members of their own community within the limits of the district in which the land is located with prior permission. For more detail on CNT Act cases, see http://jhr.nic.in/hcjudge/data/50-2313-2008-25102008.pdf

4 The Fifth Schedule provides for special legislative powers to the governor, the governor's report, and most importantly the tribal advisory council. It worked to promote and protect distinct traditions, customs, culture as well as to ensure overall ethno-development of the tribal people in India.

5 Benign integration refers to the continuing policies of protection, exclusion, and positive discrimination.

6 https://www.bloombergquint.com/business/2017/02/17/jharkhand-signs-209-mous-worth-rs-3-lakh-crore-in-investment-summit

7 Innis, a noted geographer who analyses the development of capitalism in Canada, has developed vast ranges of specialized vocabulary of political economy to analyse and discuss the laws and relations of capitalist development and its discontents. His keywords are elasticity, unused capacity, incidental price system, fixed overhead costs, disturbance, monopoly, factors of development, disequilibrium, business cycle, geographic unity, capital movements, cyclonics, liquidity preference, vested interests, and technique of production. For more discussion, see (Drache 1982) https://journals.uvic.ca/index.php/ctheory/article/viewFile/13914/4688

8 For more information, see, https://www.avenuemail.in/region/kolhan-commissioner-directs-closure-mining-crushers-brick-kilns/95657/

9 http://ismenvis.nic.in/LatestNewsArchieve.aspx?Id=14238&Year=2017

10 https://www.telegraphindia.com/1150119/jsp/jharkhand/story_9131.jsp

11 https://www.firstpost.com/india/illegal-mining-tenure-of-shah-commission-cannot-be-extended-centre-tells-sc-1359389.html

12 For more, see, http://www.indiaenvironmentportal.org.in/files/file/first%20report%20on%20illegal%20mining%20of%20jharkhand%20vol.%202.pdf

13 https://economictimes.indiatimes.com/industry/indl-goods/svs/metals-mining/crores-lost-in-illegal-mining-and-exports-in-jharkhand-goa-mb-shah/articleshow/39636510.cm

14 The case was also published in his book. See (S. Roy, 2012, pp. 284–288) http://www.saryuroy.in/books/16_38_02_jharkhand_ka_madhu_kodha_loot_raj.pdf

15 For more details, see http://www.mondaq.com/india/x/491966/Mining/The+Mines+and+Minerals+Development+and+Regulation+Amendment+Act+2016+2016+Amendment

16 For more, see http://www.thehindu.com/opinion/lead/of-mines-minerals-and-tribal-rights/article3419034.ece

17 Hold strategy refers to distribution of money to officials, local leaders, or influential persons to get legitimacy and address the oppositional force. For more discussion, see https://fas.org/irp/doddir/army/fm3-24.pdf

18 Cited in (Ashokan & Rangarajan, 2011)

19 For a discussion on tribal religions in India as one essentially revolving around a spirit-nature-man complex. For more, see (Mann, 1981).

20 In casino capitalism, there is no factual connection between the effort (investment) and reward (result). For more, see the work of (Strange, 1986)

21 We will sacrifice our life but will not give an inch of land to them. In every protest against land lease and extractive industry, Ho people often raise slogans at the time of protest against state or any company (*Jaan degai magar jameen nahi degai*).
22 Advocacy politics is a direct and practical effort to encourage and contest policies and authorities and even the system and manner in which resources are produced and distributed.
23 This incident happened in 2015. Villagers had no knowledge about the orders of the government officials.
24 *Adivasiyat* is a way of life. It counters the ideological domination of capitalism, as already discussed in Chapter 3.
25 Government (three forms of judiciary, executive, and legislative).
26 For common sense, see (Krancberg, 1986) and (Hoare & Nowell Smith, 1971).

Chapter 6

The State in the Making

> Modern science, which displaced and replaced God, removed that obstacle limits on freedom. It also created a vacancy: the office of the supreme legislator-cum-manager, of the designer and administrator of the world order, was now horrifyingly empty. It had to be filled or else … . The emptiness of the throne was throughout the modern era a standing and tempting invitation to visionaries and adventurers. The dream of an all-embracing order and harmony remained as vivid as ever, and it seemed now closer than ever, more than ever within human reach. It was now up to mortal earthlings to bring it about and to secure its ascendancy.[1]
>
> – Zygmunt Bauman (Modernity and the Holocaust)

6.1 Introduction

The supreme administrative and legislative authority of the new world is the State which has tried to address the consequences of development as an immanent process (Cowen & Shenton, 1996) by setting new high modernist projects and top-down schemes that can be conceptualized as intentional projects (Cowen & Shenton, 1996) in everyday making of the State. Additionally, it enhances the enabling condition for the global capitalist market to make people more dependent on the market through the State polity. The post-colonial India has adopted the policy of integration to deal with the indigenous peoples. Adivasis' self-hood, ethnicity, historical myth, self-rule, autonomy, and connection to territory led to Adivasi identity and resistance politics resurgence. These politics became a part of a more extensive counter-hegemonic process that questioned and challenged State capitalism on the question of access and control over productive resources (Heredia, 2002). In contrast, their demands and political expressions have been linked with the idea of customs, culture, and place that is also acknowledged by the Indian constitution under several acts and provisions. It shows how State's policies work in Adivasi region where the process of development guided by the principle of primitive accumulation also dispossessed the community who are supposed to have control over their land and resources. Control over land

DOI: 10.4324/9781032637990-6

and resources is an essential part of their sacral and alternative imagination against the State mode of development. To incorporate Adivasi agendas and politics into mainstream neoliberalism, Prasad (2016) argues that the State promoted the ideal of self-rule and management of resources. One of the integration methods, called State formation, is supposed to have provided the decisive sociopolitical framework for transforming the sacral polity of the Adivasi system through participation in the election, development projects, vocational training, etc. (Xaxa, 1999, p. 1520). Contesting such claims, scholars like Shah (2010) have stated that the tribes in Jharkhand participated in State-conducted elections to keep the State away to preserve their customary institutions through sacral politics. Additionally, Chandra (2013a) argues that the State is not a non-existent entity; rather, it is a political State that provides a space for the Adivasis to claim their rights in the form of negotiation, resistance, or voice. Negotiations with the State are also negotiations of State power from below. Adivasi politics vis-à-vis the post-colonial Indian States are characterized by negotiating claims that seek to ameliorate socio-economic conditions as well as to alter State–society relations in legally demarcated Schedule V areas. Adivasi–State relations in contemporary India are amicable (Chandra, 2013a, 2013b). No one can deny the fact that Adivasis have also selectively got benefits from the State in terms of job, reservation, and education but this has also degraded the self-determination and power of the Adivasis to control their condition of existences, fate, and labour. It is because

> people encounter development from their mundane, daily concerns to build and improve their livelihoods, to build places they enjoy being in, to give meaning to their lives through these livelihoods and places, and to maintain, and as, far as possible, to extend the degree to which they can exercise control over their mode of existence.
>
> (Bebbington, 2000, p. 513)

They are finding new ways to control the elements of development through local action, resistance, and participation.

The Ho mode of existence is embedded into their social system, but to integrate the Adivasi region into the national economy, the State has chosen the path of economic integration. This integration seems to proceed through the interrelated change in both production and market relation. The policy of economic integration in the form of social formation has shaped and restructured the economic and sociopolitical system of the Adivasis. The socio-economic system of the Adivasis has moved from the one that is based on 'stability and self-consumption' to one based on 'accumulation and market orientation' (Nathan & Kelker, 2003, p. 1955). Thus, it demands an inquiry to understand how institutionalized and power asymmetry in resource management have led to a new kind of people with their emerging values, definitions, and understanding of life worlds. This chapter shows how new environmental actions and shadowy regulatory systems lead to a new kind of people and arrangement. In 1996, Panchayati Raj was introduced as a policy initiative for decentralization in India. The 73rd amendment of Indian Constitution in 1993 granted constitutional status to three-tier local elected bodies or panchayats to strengthen participatory democracy. In this regard, the State promoted decentralization and people's participation in decision-making through decentralized democratic institutions at the village level to manage natural resources. To give rights to the Adivasis over their territory and control over natural resources, the Panchayats Extension to Scheduled Areas Act (PESA) was passed by India's Parliament in 1996. Prabhu (2004) writes that 'PESA moved from delivery of development to empowerment; from implementation to planning; from circumscribed involvement to conscious participation'. Cooperation, participation, and interrelation between different institutions are crucial in the decentralization and efficient management of resources. Decentralization is the 'relocation of administrative functions away from a central location to lower and more localized levels of government; devolution is the relocation of power away from a central location' (Enters, Durst, & Victor, 2000) and leads to more efficient, sustainable, and equitable outcomes (Lele, 2004). Literature on decentralization and the State building context in resource-conflict areas shows that functional democratic institutions are scarce (Anten, 2009; Bergh, 2004; Brinkerhoff, 2005).

Additionally, Sareen's (2015) study shows that democratic institutions can be functional in resource conflict areas. Still, his study was mainly based on the functioning of the Gram Sabha in the context of forest regulation and MGNREGA Schemes. The study shows that the effects of decentralization have enhanced the State capacities among the Ho Adivasis. Decentralization is about bringing the State back in the picture of the Adivasis positively and legitimately, but in the case of West Singhbhum decentralization is in compliance with many institutions, associations, and agencies along with some non-democratic political association that worked only to enhance the State building in a twilight democracy. This book intends to show, in West Singhbhum where extractive projects are at peak, that the democratization of governance and equitable outcomes are rarely practised and achieved at the time of delivery of developmental schemes or at the time of establishment of mining operation.

Additionally, it enhances commodity consciousness among the Adivasis and produces inequality, injustice, and the class of political elites and fixers (middle man). Political fixers, middle men, and brokers access the State's top-down prioritized projects and policies in the name of participatory resource governance. At the local level, policies and schemes came from the top-down and local people, whereas at the grassroots level it is said that the local villagers frame policies by neglecting the political representation of the Ho Adivasi. It includes people's opportunity to influence decision-making related to environmental rights and social distribution of resources (Schlosberg, 2007). Participation is a kind of engaging people into devolution and State-guided decentralization. As shadow practices, it became a feature of the making of the everyday State in Singhbhum where local Adivasis have experienced it on a daily basis. As a process, it is done in the name of democratization of governance that gradually integrates the sacral polity, ideas, values, and institution of Adivasis and makes the Adivasis citizens without an alternate imagination and sacral polity. These complex processes help in strengthening the State capacity and presence. The State schemes that are meant to enhance the livelihoods of the local Adivasis substantively produced political patronage rather than achieving its objectives.

Here another question arises as to how resource governance and participation, recognition and capabilities of the Adivasis are addressed at the time when discursive power of State is engaged in facilitating extractive projects and other developmental schemes are functioning to address the demands of surplus people who are being dispossessed due to transformation of ecology and erosion of choice (Li, 2014). Taking the State as a monolithic entity and the local people as a homogeneous class having embedded resource interests is an oversimplification especially within the context of development-induced dispossession or of the people as beneficiaries of State services and improvement schemes (Agrawal, 2001, 2005). The State formulates people's ways of thinking about themselves and their relationships with each other through these schemes by employing its power and apparatus. Power, here, can be defined as the capacity to affect the outcome of decision-making processes and implies a genuine role in decision-making. Power as a strategic game signifies structuring the possible field of action of the subjects. It can be in multiple forms, e.g., ideological manipulation or rational argumentation, honest advice, and economic exploitation (Lemke, 2010).

In the case of environmental clearance to mining industries to employment scheme implementation, Gram Sabha, Panchayati Raj institutions (PRIs), and other institutions served as a tool for the State. State schemes and ideological projects' compliance with the extractive projects can be expressed as the State's bio-politics that can be translated into the State's desire to control the lives of the Adivasis and making them what Foucault calls 'subject of needs, but the object in the hands of the government' (Faubion, 2000, p. 12). These schemes and projects are a political step because they undermine the issues of justices in terms of participation, recognition, capabilities, and distribution (Schlosberg, 2007). The threat of environmental justice dimensions is common in 'hyphenated society'.[2]

Adivasis' relation with the State can be conceptualized as obstructing the State (Gell, 1997) and keeping the State away (Shah, 2007). As a signifier, it showed how their societal relationship with the State started changing over time through everyday negotiation and accommodation to strengthen State building. State building is about controlling violence, establishing legitimacy, and building capable and responsive institutions. These are all long-term and possibly conflict-ridden processes. State building is a leading priority in fragile and resource conflict settings, but ongoing State building challenges persist in States under comparatively more 'normal' development settings. The discussion about the State's functioning is originally an ideological one, focused on how the State claims legitimacy and establishes a monopoly of governance in the name of the constitution and the rule of law. These laws and provisions govern the Adivasis, who are intimate with nature and thus interlinked with the theme of environmental justice. State officials and developmental professionals discussed the need to participate in development schemes and projects. State-promoted participatory governance through PESA and other acts requires meticulous understanding of the process of State building that produces and legitimizes misdistribution practices, including cultural recognition. This is because misrecognition is an institutional process in development schemes and projects that devalue community, people, and place in comparison to others. It is because structural asymmetries and psychological components of misrecognition are located in any distributional schemes and institutional arrangement of the State which is a kind of injustice.

6.2 Analytical and Conceptual Framework: Justice, Democracy, Livelihoods, and Assimilation

State developmental projects mainly expand the commodity frontiers through 'accumulation by dispossession', co-optation, and assimilation. It affects the people's identity and culture (Fraser, 1998) and their ability to participate in the decision on developmental projects and livelihood schemes that may affect their life worlds. State's decentralized schemes also undermine the ability and capabilities to develop the notion of freedom Sen (1999). The State promoted extractive industries and highly modernized schemes that affect access to the natural resources of the natives. Access to land, forests, and other resources is the base for their livelihoods and culture. This access is hindered either by dispossession or by high modernist schemes or the ecology transformation that sustains the people's livelihoods (Bebbington, 2000). 'People's access to resources or practices to engage with the environment have become so altered that it can be termed structural displacement' (Oliver-Smith, 2006, p. 155).

The issues of development-induced dispossession, resistance, and transformation of the Nehruvian State to a neoliberal one have facilitated several schemes and rearranged the policy of resource allocation and distribution (Kumar & Mishra, 2017). The empirical analysis carried out in this chapter is built upon the political ecology of the State-driven intervention, issue of governance, access to resources, erosion of choice, and justice. State formation requires the incorporation of the local people into the modern process of development that induces assimilation. The term 'assimilation' refers to the practice of incorporation, which denies differences in culture and value (Escobar, 2006a). Assimilation entails how the local community is forced to choose livelihood and other opportunities provided by the State, companies, and several developmental agencies. It also breaks the complex articulation of the multiple dimension of livelihoods (Bebbington, 1999, 2012, 2013). However, any form of assimilation is bound to erase the local identity.

When State extractive projects, schemes, and policies threaten the livelihoods of the local Adivasis, resistance may emerge in the defence of it. Still, the discursive practices of State's 'will to improve' (Li, 2007) colonized social reality, undermining the multiple dimensions of justice (Schlosberg, 2007). These dimensions of justice – distribution, recognition, participation, and capabilities (Schlosberg 2007) – deal with human and community rights over territory. These dimensions are also included in the constitutional framework of India under the provision of PESA, CNT, and other safeguards provisions in general and Ho Adivasi in particular. It also deals with the social, economic, cultural, and territorial rights as the right of the Adivasi and demands for procedural justice and recognition and participatory justice (consultation, free, prior, and informed participation) before or at the time of the scheme and project implementation. The political ecology approach helps us to understand how natural resource territory is integrated into the State economy through different acts and policies that claim that it is meant for the empowerment of the local people.

6.3 Why Government Schemes?

Newly created States in the independent nation are subjected to many external and internal factors, of which the market-driven factor leads over others. Jharkhand is in the process of State building, which requires amending steps for inclusive politics through struggles over citizenship and rule of law (Sundar, 2009a). The inclusive policies and politics are to address the resistance and encourage the Hos' participation in participatory electoral democracy. Various schemes and policies of the State aim to facilitate the extractive projects, and the market represents a symbolic representation of the State's modernity. The failed project could be seen as well planned and aesthetic, which is central to high modernism. Authorizing access to natural resources and State schemes empowers local people to have resultant co-optation and assimilation. The policies and livelihood scheme of democratic decentralization enable the top-down authorization of the State, where the presence of the State is experienced through different bureaucratic levels (Sareen, 2015; Sareen & Nathan, 2018). A State is a facilitator and critical actor for the State building and overall developmental process. It implements and involves investment-friendly projects through expropriation and eminent domain. The State's blending of regulatory simplification, sovereignty, and coercion eases the entry of capital into strange territories that usually lack pre-existing formal land rights or include abundant forms of communal tenure (Borras, Franco, Gomez, Kay, & Spoor, 2012, p. 858). However, the institutional implications are fundamentally different from previous formations of the enclosure. In conflict and fragile situations, the State usually disburses its sovereignty by creating schemes that lessen the magnitude of local people's resistance, forcing them to surrender their normative and economic rights to the capitalists.

As a part of State intervention, these schemes provide enabling conditions of governance needed when the State facilitates extractive industry. The State shows that the decentralized policies and programme are meant for the local community and can be conceptualized as the State's 'will to improve' the commons. The multi-developmental discourse and scheme of the State operation in a particular place or territory ensure that local communities will benefit from State neoliberal capitalist projects from the beginning and throughout their life. Ghani, Lockhart, and Carnahan (2005), for example, argued that the

> State has ten primary functions: viz, the legitimate monopoly on the means of violence; administrative control; management of public finances; investment in human capital; delineation of citizenship rights and duties; provision of infrastructure services; formation of the market; management of the State's assets (including the environment, natural resources, and cultural assets); international relations (including entering into international contracts and public borrowing); and the rule of law.
>
> (p. 9)

The State does it through the executive and legislative body called the government. Foucault (1991) described it as the 'emergence of government', that is, grounded for improving the condition of the community and population through the deployment of bio-political programmes. The State's discursive practice that keeps non-capitalist economies and institutions in an economy gradually devalued them and confined them to the margins. Harriss-White (2014) identifies household and self-employment as sites of non-capitalist production where State schemes work. To govern through communities and make them responsible for their own fate is a feature of State schemes that can be easily observed, especially in employment schemes, projects, and corporate social responsibility schemes. This neoliberal intervention through schemes and projects works as a safety net for the State and corporate to address the Adivasi resistance through Ponzi schemes (Harvey, 2007) or failed schemes (Kumar & Corbridge, 2002; Scott, 1998a). Scott has emphasized that the State's attempts towards ecological simplification and legibility, along with the complex dynamics whereby State's expertise often overrides other's knowledge, has resulted in the failure of the policy (Harris, 2017).

The nature of these schemes and participatory programs exemplifies the principle of 'creative destruction' (Harvey, 2007). It has a mechanism that demands that the poor, or once dispossessed, supply their own 'employment as entrepreneurs' (Elyachar, 2005) or try to get reabsorbed in their subsistence economy through State policies and schemes of employment. In brief, it can be said that State schemes of employment are used for absorbing the surplus population (Li, 2012) who have lost their ecological base of livelihood (Blaser, Feit & McRae, 2004; Parajuli & Kothari, 1999), due to 'second contradiction of capitalism' (Lane & O'Connor, 1995; Stroshane, 1997), and have been accompanied by a development of the exploitative relations of capitalist accumulation. Post-colonial Indian capitalism was adopted at dispossessing large masses of people from their means of production under the rubric of capitalist industrialization, owing to the capital-intensive nature and the slow rate of investment. A large proportion of the dispossessed due to developmental projects were unable to find gainful employment as wage-labour within the formal industrial sector. Therefore, unlike developed countries where primitive accumulation is accompanied by capitalist accumulation and the absorption of labour, in post-colonial India the dispossessed simply become a large mass of 'surplus' labour (Li, 2012). These labourers are alienated from their means of production and excluded from capitalist class relations. Sanyal (2007) argued that these labourers were unable to become wage workers and became a 'marginalised of the wasteland'.[3]

In West Singhbhum, through various schemes, programmes, and facilitation of extractive industries, the State has imposed its own will on the Ho community to address the issues of dispossession, conflict, and livelihood. State also reinforced several schemes through politics of decentralization to shape environments for human action and interaction (Paulson & Gezon, 2005), which include technologies of rule (Lemm & Vatter, 2014) and strategies of the invisibility (Hammond, 2011) that strengthen the State building programmes. Hence, the will to improve the Ho community through local-level resource governance, management, and schemes is a useful approach of political ecology to understand the gap between the attempts and accomplishment in State intervention.

6.4 Politics of Decentralization, Legality, and Shadowy Practice of State

Post-colonial Nehruvian Indian State retained the customary institutions of the regions by recognizing colonial laws like Wilkinson Rule, 1837, Chotanagpur Tenancy Act (CNTA), 1908, and provisions under the Scheduled V areas. State also established statutory panchayats in the scheduled areas of Jharkhand, but Adivasis' participation in the elections, decision-making, and resource governance was minimal in the statutory panchayat.[4] As a result of Jharkhand's acts on panchayats, most of the Gram panchayat's heads were non-tribal or other than the customary head. It induced the conflict between the power of the customary institutions and State-made democratic institutions. Troisi says that 'the traditional councils were considered as a means of fostering group consciousness, the new government institutions are looked upon as a force that is disintegrating the tribal village solidarity' (Troisi, 1978, p. 154). This has given a new political arrangement for the State to deal with the institutions, and it continued till the formation of the provision of the Panchayati Raj Schedule Areas Acts of 1996, which talked about the retaining of customary self-governing institutions and mandated the bestowing upon of several powers on Gram Sabha, e.g. decision on land alienation, management of minor forest produce, minor mining leases, maintenance of water bodies, and so on. On the other side, Jharkhand, after the issues of a long ethnic conflict for control over the resources, has manipulated the Central PESA law through a bill passed in the State legislative assembly called Jharkhand Panchayati Raj Act (JPRA; 2001). Initially, it was widely protested by the non-Adivasi association because there was no seat granted under reservation policy in Gram Panchayat, Zila Parishad, and Panchayat Samiti for the non-Adivasis. As a result, non-Adivasi organizations, like Chatra Yuva Sangharsh Samiti and the Jharkhand Pradesh Panchayati Raj Adhikar Manch (JPPAM), deriving support from OBCs (mainly Kurmi-Mahato) argued that reserving the chair seat for the Adivasis in scheduled areas violates their citizenship rights. They demanded to 'deschedule certain areas where they claim Adivasis are in the minority' (Sundar, 2005, p. 4430), and, as a result, the State finally reserved the post of upmukhiya (vice mukhiya) in Panchayat Samiti and vice president at Zila Parishad on 10 October 2003 for the non-Adivasis. It showed how Jharkhand State favoured other pressure caste group institutions against the provision of PESA, which demands retaining the customary governance system. While Sections 4 e and f and Section 4(m) (vii) of the PESA empower the community to take charge of development and integrate progress in their culture and heritage, they also ensure that 'planning is geared to meet felt needs, programs reach the real beneficiaries, the process remains transparent and accountable' (Dobhal, 2011, p. 18). Since 1996, in scheduled areas, then, India's constitution states that the village assembly (Gram Sabha) can 'safeguard and preserve ... community resources and the customary mode of dispute resolution'.[5] Village assembly can exercise authority over natural resources (land, forest, water) in accordance with traditions (Dandekar & Chitrangada, 2011). In addition, it formally empowered the village assembly to control and implement local development plans and projects and to select beneficiaries under development schemes (GoI, 1996: clauses 4e, m). The approval of the village assembly is needed for all development works and projects, including certification of work executed correctly by higher level institutions like PRIs or government line agencies. But overlapping legislation allows a government agency, mainly the mining department and pollution clearance board, to favour a

Table 6.1 Review of PESA in Jharkhand

State	*Land Acquisition*	*Excise*	*Mines and Mineral*	*Forest Product*	*Agri Produce Market*	*Money Lending*
Jharkhand	No	No	No	Yes[1]	No	No

Source: Author

Notes:
1 Jharkhand government adopted a resolution on 8.2.2007 conferring ownership rights of MFP to GP. For details, see http://pesadarpan.gov.in/documents/30080/45009/Jharkhand.pdf/7c90a28a-7b6b-43d0-be86-6834c42937bf

mining company. It is very common in Jharkhand. The Jharkhand government manipulated the Central PESA Act and its provision to facilitate mining projects and ideological base of capitalism through schemes and co-optation. The Ministry of Panchayati Raj of India has reviewed the status of Central PESA and found that the Government of Jharkhand has not framed the PESA rule in a real manner.

Table 6.2 shows that Jharkhand is meant for the tribal's self-rule, autonomy, and their control over the resources and territory is not recognized by the State. The State has only recognized some customary powers of the Gram Sabha to show that the State is with the Hos and their institutions. In the case of resource control, planning, money lending, land alienation, mining lease permission, and land acquisition, Jharkhand has not given several vital powers to Gram Sabha. Devendra Nath Champiaa, a former member of the Legislative Assembly and vice speaker of the Bihar Legislative assembly residing in Chaibasa, mentioned that PESA is internally contradictory because it allows both elections and customs; and the newly passed JPRA has also violated the clause 4 of the PESA.[6] He also filed a public interest litigation (PIL) against the panchayat election in scheduled areas and violation of PESA in Jharkhand State through JPRA, but the State defended its right to pass the act.[7] The most exciting thing that came to light after reading the judgement given by Chief Justice Altamas Kabir was that neither the State nor the party who filed the PIL talked about the control over the resources, land acquisition, and consultation before the project implementation. However, Justice Kabir also spoke about the administration of Kolhan (Singhbhum). He said that the district administration must fully follow the Manki and Munda system. However, the aspects of the Kolhan administration are handling civil cases in the areas under the Wilkinson Rule.

Tahsildar is an agent of Manki to assist the collection. They all work on a commission basis and keep a certain percentage of the rent collected. The revenue is collected through the Manki Munda system and the implementation of the provisions of the Chotanagpur Tenancy Act, to prevent alienation of the lands of the Adivasis.[8] Recognition of Manki Munda is necessary to access and control the region. They represented the philosophy of sacral

Table 6.2 Status of Tribal Customary Law Over Resources

Sl. No	*Provisions of the Act Status*	*Status*
1	4(d) Customary mode of conflict resolution by the Gram Sabha	yes
2	4l: Selection of programme beneficiaries by Gram Sabha	Yes
3	4(f): Gram Panchayat to obtain utilization certificate from Gram Sabha	Yes
4	4(h): Nomination by State government of persons of ST not represented in intermediate and district PRIs	Yes
5	4(i): Consultation with Gram Sabha or PRI before land acquisition and resettlement and rehabilitation	No
6	4(j): Planning and management of water bodies by Gram Sabha or PRI	No
7	4(k): Recommendation by Gram Sabha or PRI before grant of prospecting license or mining lease	No
8	4(l): Recommendation by GS or PRI before exploitation of minor minerals	No
9	4(m)(i): Power to restrict sale of intoxicant to PRI and Gram Sabha	No
10	4(m)(ii): Ownership of minor forest produce to PRI and Gram Sabha	No
11	4(m)(iii): Power to prevent land alienation to PRI and Gram Sabha	No
12	4(m)(iv) Power to manage village markets to PRI and Gram Sabha	Yes
13	4(m)(v): Control money lending to PRI and Gram Sabha	No
14	4(m)(vi): Control of social sector institutions and functionaries to PRI and Gram Sabha	Yes

Source: Author

polity – a sacral polity embodying values such as egalitarianism, consensus, and the dispute resolved within the boundary of the village through open discussion and negotiation, but in recent times there are many vacant posts of Manki and Munda due to the State's unwillingness in appointing them.

Now the question arises: How does the State implement the schemes in these regions where there is a vacant post of Manki Munda? This post was necessary for the Kolhan administration because they collected the revenue for the State, handled the civil dispute within the village's boundaries, and reported to district authorities. Now without them, how does the Gram Sabha conduct a meeting, who does preside over it, and how can State developmental policies and schemes reach the locals? There are many questions related to the legitimacy of Kolhan administration in these villages where the traditional chiefdom (Manki Munda) post is vacant and not recruited or approved by the district collectors. State officials do not have an appropriate answer to this question. A PRI officer of the district said that he was newly appointed to the district and did work on the matter. During the

Table 6.3 Manki Munda Filling Up Positions in West Singhbhum

S.no	*Block Name*	*Manki Post*			*Munda Post*		
		Post Sanctioned	*Working*	*To Be Filled*	*Post Sanctioned*	*Working*	*To Be Filled*
1	Manoharpur	4	2	2	76	70	6
2	Sonua	5	5	0	97	65	32
3	Goilkera	5	4	1	117	101	16
4	Gudri	2	2	0	30	30	0
5	Badgoan	9	5	4	94	56	38
6	Chakadarpur	5	5	0	59	52	7
7	Anandpur	0	0	0	23	0	23
8	Sadar Chaibasa	6	5	1	72	68	4
9	Manjhari	6	6	0	64	50	14
10	Tatanagar	7	7	0	76	67	9
11	Jhinkpani	5	2	3	51	45	6
12	Hat Gamharia	4	4	0	54	44	10
13	Tonto	3	2	1	77	60	17
14	Khuntpani	11	11	0	115	94	21
15	Majhgaon	7	7	0	88	72	16
16	Kumardungi	8	5	3	83	66	17
17	Noamundi	5	3	2	69	53	16
18	Jagannathpur	8	6	2	89	70	19
	Total	100	81	19	1334	1063	271

Source: Author

conversation, he said, 'West Singhbhum is scheduled district, therefore, Munda presides over the meeting of Gram Sabha, and we (government agencies of the district) organized training for the Mundas to be more efficient in their work' (personal interview, 9 September 2016).

Mobilization for self-governance is a State agenda but is based on the secular kind of alternative government in which the Adivasi cultural beliefs and religion have no space (Shah, 2010). It is promoted within the framework of a post-colonial capitalist State where the economy determines the validity of the institution. Shah (2007) mentions that Parha fairs among Munda tribe were promoted as a traditional authentic polity with sacral connotation against the corrupt State. Still, in case of the Hos, it is observed that the Manki Munda system also works with traditional authentic polity but is limited only to customary disputes. The State is making them more accountable by giving them training and treating them as State agents. The continuous training and meeting determines their accountability towards the power, which gradually degrades the alternative imagination and polity of the Manki Munda system. Since the making of Jharkhand, poverty alleviation programmes and employment schemes have been started. These schemes tied up with the 'ideology of national development constituted the foundation of the newly independent Indian State, and development planning was the most important instrument adopted for this purpose' (Chatterjee, 1998, p. 86). Notably, however, the State attempted to segregate politics from (planning) development or depoliticization. Many prominent scholars have observed the nature of the 'accommodative politics' that arose within the first decade after independence and continues till today in the neoliberal fabric (Bardhan, 1984; Kaviraj, 1988; Kohli, 2014; Rudolph & Rudolph, 1987). The accommodative politics and planning started in the year 1955 through a project of community development, the slogan of community development in the Bihar Panchayat Raj Act was 'service through guidance and self-help'. It was the direct intervention of the State in the region, which defined that the people of this region are traditional and in need of development. To describe Hos' social world and territory, the State deploys the power and discourse of development to rearrange people, territory, and institutions and creates regulating mechanisms. It started from the community development schemes through the Panchayati Raj system. The traditional chiefdom system had sharply clashed with the new Panchayati Raj system. By 1955, new blocks were created for the development of the region. The territorialization projects of the State gave full power to the district collector. The post of district development officers was changed to block development officers. The block development officers were also circle officers. Under him, there were other line agencies like agricultural, cooperative, horticulture, officers etc., followed by Gram Sevaks (village-level workers) and others. A village-level worker's (VLW) job was to follow the instructions of State bureaucracy and serve as a bridge between village and State officials at block level. Earlier a village-level worker acted as supervisor, and he also has a political involvement and responsibility towards village Panchayat and administrative responsibility towards block developmental officers.

The panchayat election of 2010 in Jharkhand was conducted under the guidelines of JPRA. As amended in 2010, Section 21 of the JPRA reserves 50 per cent of seats for women.[9] It was also an opportunity for some Mundas who were in quest to gain power through the State's idea of democracy and for the political party to strengthen its network in the remote areas of the Singhbhum. Manki and Munda, who opposed the initial election, started negotiating with the political parties. In a meeting dated 6 January 2011, Ho Samaj Maha Sabha officially revealed that the tribal outfits were not willing to field any candidate of their own and would back the candidate of a political party that supported the cause of the Adivasis. In a meeting with one of the members of the Ho Samaj Mahasabha, he said that they had a precondition that they would support that party who would implement the CNT Act and stop Adivasi alienation from their territory (personal meeting dated 12 January 2016). The rural poll encountered significant opposition from various customary and political associations of the Ho Adivasis, including Ho Mahasabha and

Kolhan Rakhsha Sangh. They came united to protest the Panchayat election. There is no definite period to restrict the compliance of the local Manki Munda system with the State agencies. Still, it is a fact that decentralization and Panchayati elections have affected the traditional local system. In one case from the field, the local Munda of Gadahatu said that the State had reserved 50 per cent seats for women in the PRI. He saw it as an opportunity and made his wife contest for the election for mukhiya. His wife filed a nomination and fought the election and won too. He said his village's *killis* (clan) are majority in number, which led his wife to win the election. But, though his wife holds the mukhiya post, he does all the work of mukhiya for his wife at the block level (personal interview 18 February 2016). Elections divided the Ho community within the clan and lineage network and provided them a sense of heterogeneity within the homogeneity of a community.

There are many cases that were observed during the fieldwork that revealed that local Manki Munda refined them in the setting of political decentralization. Mundas opposing State intervention are now involved in selecting mukhiyas, zila panchayat president, and other posts to secure their legitimacy through lineage network. The collaboration and nexus also co-opted the local Manki Munda system to promote the idea of the State (governance) in day-to-day activities. Several political parties, an essential part of State democracy, recognized them as political leaders instead of the localized subject of moral economy. That has resulted in Hos starting to engage with the State through mundas and mukhiyas. The Jharkhand government also started to give Rs 3,000 monthly as an honorary amount and a laptop to village Munda. Earlier, Munda used to collect the revenue manually and deposit it to district administrators. The State started to give training and organized episodic meetings with government officials to make them more accountable to the State. It is because, until today, government line agencies lack the strength to deal with the Adivasi polity.

By educating Manki Munda about the modern notion of progress, the State started to impose its assimilationist ideology. This is a kind of fraudulent way to educate one for their benefit, which can be observed in State practices as ambivalence (Sundar, 2005, 2009b). In Adivasi regions, this has mostly been taking place through processes of subjugation co-option of the chieftains and indoctrination into State political system which involves resource appropriation, coercion, and accumulation (Puthumattathil, 2014; Shah, 2010; Sharma, 1959, 2007) that restructure the local class structure and the notion of traditional polity. After the formation of Jharkhand, one of the initiatives by the State was the Manki Munda fund. During a conversation with the Manki of Chadhaipir, he said that the fund amounting to Rs 50,000was given to the Munda of each village for two years. The State directed Munda through district administration to spend this amount to develop their *hatu* (village). Still, Munda, unaware of the State polity, failed to spend this amount. In many places, the Mundas were manipulated by the block officials into not spending the fund. After a period of two years, it was observed that in many villages the fund was not properly utilized. It happened because no proper instructions came from the State at district and block administrations. So the fund was not granted following that year, and an allegation was made against the local Mundas that they were incapable of handling the development schemes. This strengthened the demand of the Panchayati Raj system to implement the State's official development schemes. Many Mundas are in favour of starting the State fund again, through which they will also be given some responsibility, like that of a mukhiya, to implement the development programmes (personal communication, 12 October 2015). It shows how the State employed its apparatus and knowledge to show that local customary leaders are not capable of developing the region without the help of the State.

6.5 Decentralization and Undemocratic Practices in the Name of Development

The State has a 'double nature as an idea and an apparatus' (Lund, 2008, p. 181). It comprises multiple governmental and non-governmental politico-legal institutions exercising authority at the local level (Sareen, 2015). This section will exemplify the 20-point programmes to reflect on the politics of the State to govern the region. The structure of 20-point programmes at the district level included a foundation of many schemes and programmes for Adivasi development. It followed a pyramid shape. One minister of Jharkhand State headed the programme of the district. A draft of targeted policies for the poor was introduced. West Singhbhum is also not an exception to this. The introduction of populist schemes and top-down targeted schemes funded by the World Bank and the central government for Adivasis served an opportunity for the expansion of local-level patronage networks between the State leaders (Politicians) and its cadre. It has created a 'new' local undemocratic institution that can be shored up by Ferguson's discourse of 'depoliticization' (Ferguson, 1994; Li, 2005a). It can be seen as a participatory development to monitor whether the government schemes of the State have technocratic rationality. In Jharkhand, 20-point programme implementation committees abound with elites, supporters of extractive industries, and party candidates. The committees have three tiers – State level, district level, and block level – which are meant to monitor the implementation of 20-point programmes. Political parties ruling the State (Jharkhand) appointed a party member in the committee to monitor the implementation of 20-point programmes. Several poverty alleviation initiatives were launched during the tenure of Indira Gandhi, ranging from rural employment schemes, mid-day meals for children, and electrification (Gooptu & Parry, 2017; Vaidyanathan, 2001), and are still being facilitated by the State. Populist politics of State and State's will to improve the Adivasis through the schemes were instrumentally deployed for electoral gain, but it is translated as a participatory democracy. The targeted programme is not based on general welfarism – the Adivasi local chiefdoms were cast as clients of the bureaucratic government. The State asked them to make a plan for the village, but the sketches, directions, and blueprints of plans and policies were determined at the top level. PRIs were cast as clients of the bureaucratic government through the extension of the targeted policies such as Dhoba, toilet scheme, housing scheme such as Pradhan Mantri Awas Yojna, etc. These programmes are implemented and monitored by the political party committees. It vastly extends the patronage politics, and through this the neoliberal State has been reaching down to the local level as a neoliberal activity and process. The dispossessed people are now visible to showcase democratized capitalism with Indian characteristics. Some of them are used as vehicles for bestowing patronage and expanding the ideas of top-down high modernism democracy, among others. The committees are packed with the outsiders, *dikus*, or the people who have lost their sacral imagination. Nominated persons are also not accountable to the local representatives and Adivasis' customary institution. Still, it has created a tool of surveillance for the State, a feature of neoliberal governmentality (Lemm & Vatter, 2014).

In the 20-point programme, it is observed, during fieldwork, that political leaders (ruling party) dispense political patronage via State jobs and schemes. Political patronage via State jobs is not only to control or surveillance but also to enhance State positions. The representative of political parties employs these selective incentives and recruits their party members in any committee to monitor State plans and policies operationalized at the district. It helps political parties to develop clients' networks, which serves two purposes: first, it helps to maintain party ideology and electoral support; second, it mitigates the risk of resistance. Political patronage, popularly known as the political appointment system or cadre deployment, is a kind of democracy pathology. It is conceptualized as an exchange relationship where various items, goods, and accommodative services are traded between the principal and the agent. Usually these services do not follow any legal prescripts. In other words, political patronage or simply cadre deployment is not a government-legislated policy, thus often leading to non-compliance with recruitment or human resource management laws and regulations. These members often misuse their power and sometimes work as political fixers who play a role in the fantasies of the State power (Corbridge et al., 2005; Shah, 2010).

6.6 Gram Sabha, Gram Panchayat, and Modern Fragile System of State

Government agencies lack the strength to implement their rules locally in this conflict-prone district. The government and its formal institutions do not govern alone. Hos' traditional institutions such as the Munda and Manki contribute to governance, and decision-making also occurs through informal collective platforms. Traditional institutions, local village assembly, and local panchayat representatives (*mukhiya*), along with the government's official agencies, contribute to the governing of the region as discussed earlier and in Chapter 5 that the Ho Adivasis' access to natural resources and State institutions can be understood through their varied and contextual meaningful action ranging from everyday resistance, negotiation, and cooperation to the beneficiary. At the time of conflict, villagers through Munda challenge the State's actions. This is the way through which the Adivasis resist the State and legitimize their access. It was observed that at the time of resource conflict (mining-induced ecological degradation, and protest and resistance against land acquisition), villagers through Gram Sabha under the presidentship of Munda gave a written petition to the deputy collector against the mining operation.[10] If the company succeeded in co-opting the local leader, including Munda, the resistance against the company or extractive industry came into the negotiable stage. The process of co-optation also induced assimilation. In this contemporary time, people's ecological priorities and self-definition do not precede their actions in a developmental context; they are very much part of the power-laden development activities. These development activities have politicized the local ecology. One of the cases discussed in Chapter 5 is about Nakahasa, which deals with resistance against a limestone mining company. After facing water scarcity and dust pollution in agricultural land, the local villagers forced the company to stop its mining operation. It was decided that Munda (customary head of the village) Chandra Mohan Bari was initially against the mining but later he and some elites of the village were co-opted by the industrialists' agents and State officials. They started to favour mining but the people of Nakahasa called Gram Sabha and decided to change the Munda. Munda is the traditional officer of the scheduled areas with executive and judiciary power. Local villagers approached the superintendent of Kolhan and deputy collector of the district. They wrote a petition letter to remove the Munda. The protestors had chosen Bari's cousin as the village Munda. Somehow Bari succeeded in managing the government officials, which showed how the common consent of the villagers to dismiss the Munda from his customary head position did not work. Villagers regularly complained about the bad character and his being unfit for the Munda position through paper petitions, but Bari succeeded in saving his customary headship. During a personal conversation, he said that villagers got compensation, then what was the need for resistance against the company. According to him, those who protested did so out of an act of personal revenge; therefore, they wanted him to be removed from the post of Munda. However, the villagers said he was unfit for the position, and that he did not worry about villagers (field notes). The case of Nakahasa village shows that the societal system is rapacious and corrupt and favours the powerful through undemocratic manners that help the powerful manage the local polity in an ad hoc manner. The drama has changed the solidarity of the village. It shows the structural mechanism of the democracy which overruled the democratic decision of the local people to favour the company and the co-opted Munda. But in many cases, including quartz mining in Abaru village of Khutpani, limestone mining in village of Hat Gamharia Block, and many more, villagers through Manki Munda system approached the State for negotiation or challenged the authority. Still, this did not happen at the time when the State provided schemes through an exclusive network of local uppers and 'political fixers'[11] without going through Gram Sabha.

During the year 2018–19, Jharkhand's government has decided to form Adivasi Vikas Samiti (AVS) in all the scheduled villages and Gram Vikas Samiti (GVS) in non-scheduled areas of the State. West Singhbhum is a scheduled district, so Adivasi Vikas Samiti has been established. It is made to implement the State's one-year targeted schemes like ponds, wells, and roads. It also determines that local people should take part in the State's developmental schemes. The current chief minister directed the block State coordinators from establishing the Samiti. In each Samiti, there will be 9 to 11 members. A woman will be the committee president, and young people will be the coordinators. The State directly transfers the money into their accounts. Villagers will determine what plan they need, and they will spend the money accordingly. The State also provides direct money to mukhiyas for the development of their ward. The formation of committee is for the participation of the youth and women in the decision-making process for the development of the village. In this committee, a plan is made for village development. The State will spend 80 per cent, while villagers must contribute 20 per cent through their labour. The customary heads, manki mundas and mukhiyas, will be special invitees to the committee. The panchayat sevak (village-level officer) has to get them formed in the village assembly (Gram Sabha), and it will function under the administrative and financial authority of the block development officer. Here questions arises: Why does each scheme talk about the road, tanks, ponds, and others alike: Since there are State agencies and a provision in Mahatma Gandhi National Rural Employment Guarantee Act (MNREGA) and 20-pointed programmes, then what is the need to implement the same schemes with different agencies and institutions? It makes the role of Gram Panchayats negligible in MGNREGA. It is amply clear that this policy violates the constitutional provisions of PESA and the JPRA but has strengthened State capacity and State building. The four key aspects of the State's strength are: (i) authority over territory and conflict prevention, (ii) policy autonomy and the ability to uphold the law, (iii) responsive, accountable service provision, and (iv) social learning. Lastly, by allowing citizens to become political actors in their own right, the small scale of local politics should promote social learning-by-doing, so strengthening political legitimacy, State building, and 'democratic suppleness' from the grassroots upwards. State policies and schemes are implemented by the mukhiyas and other undemocratic institutions, showing how the State intentionally degraded the right and power of the Kolhan administration (Manki Munda system). Top-down schemes produce an omnipresent idea of the State, and the production of the idea of development is mediated by the Manki Munda, local PRI, and other informal and formal institutions that show the multiple governmentalities of the State. State recognition of several institutions helps the State, in turn, to monopolize its objectives.

6.7 Tolerating Top-Down Inequality, Developmental Projects, and Injustice

Government's support to the extractive industries mediates power asymmetries. State's institutional support has greatly boosted the extractive projects and neoliberal schemes that undermined the issues of justice and increased the State's legitimacy. State bureaucratic arrangement and mounting to the industry created an imbalance in the region's ecology. The conceptualization of 'will to improve' (discussed in Chapter 3), involving 'programs to move populations from one place to another, to better provide for their needs' (Li, 2007, p. 2), is a helpful approach herein. This Foucauldian-oriented, political-ecological perspective focuses on the gaps between attempts and accomplishments in development interventions. The State's attempts to develop the local population threatened the different dimensions of justice that caused everyday resistance and protest. Still, as time went on, resistance slowly weakened and lastly got diluted after the village's co-optation of local elites and uppers.

In the village of Nakahasa, people do not want the State's penetration. Livelihood enhancement schemes like MGNREGA were also not demanded. Only two people are required to work under MGNREGA. The livelihoods of the village are multidimensional. The agricultural land, forest, and local resources are sufficient for their livelihoods. Land and local resources are imbued with cultural and political significance. Villagers don't want mining and State penetration in the name of development. An activist of Yuwa Jhumar said that villagers did not want the company to mine because they knew that if the company extracted the limestone, it would hardly give work to 10 to 15 people. Still, it would dismantle the complex embedded identity, solidarity, and freedom to pursue the lives villagers want. He cited an example of a company mining in the nearby villages of Tonto and Jhikpani block. The mining operation is very old in these villages. People initially resisted the company in these regions, but the State supported irregular transfer, allotment of agricultural land, and continuation of mining, thereby weakening their agitation. People now realize that mining resources are a curse for their region. The history of mining in this region is a history of illegalities done by the government agencies and bureaucracy to legitimate the top-down developmental model of the State, which undermines the Adivasi land, identity, and multiple dimensions of livelihoods.

Manki Tubid, a local tribal activist, says that State developmental activity threatens the local people's life worlds in three ways. First, they physically dispossessed the local people, then, second, the transformation of the local ecology followed. Third, and last, they started to provide others with means of livelihood through schemes (personal interview, 22 September 2016). His statement helps one to understand why the Hos rejected or resisted the mining or State's scheme proposals. In field areas, it was observed that in the case of minor mineral mining like limestone, people often reject the ideas of State and company proposals for the development of the region by providing employment, education, and infrastructure. Once the company starts mining, it will impact the local Ho Adivasi subsistence base. State schemes and mining proposals disarticulate the dimension of justice. Recognition, capabilities, and participation are never considered at the time of implementation of schemes and policies, and these schemes are the anti-political machines of the State (Ferguson, 1994; Scott, 1998a). They aptly argue that it helps the State to advance domination of the State over its (potentially rebellious) subjects while simultaneously rendering its political consequences unnoticed. Thus, Adivasi villagers seldom notice how a petty contractor becomes a panchayat 'representative' to see why and how they are 'seized upon, understood, reworked and possibly contested by differently placed people within the population of the poor' (Corbridge et al., 2005, p. 4). Development institutions generate their form of discourse, and this discourse simultaneously constructs [the beneficiaries of the schemes or the people who get manual work] as a particular kind of object of knowledge and creates a structure of knowledge around that object having a regular and episodic effect which includes the expansion and strengthening of technocratic State power. State, through schemes and by engaging customary institutions in bureaucratic set-up, succeeded in delivering that local Hos deny the developmental activities of the State, and with that State succeeded in suspending its political effects in the form of resistance and governance Development is an 'anti-politics machine' – depoliticizing everything it touches, 'everywhere whisking political realities out of sight, all the while performing, almost unnoticed its own pre-eminently political operation of expanding bureaucratic state power' (Ferguson, 1990., p. XV).

As a form of an anti-politic machine, the developmental projects and schemes of West Singhbhum have a history of illegalities and injustice. Any mining industry was established through the State's 'will to develop the region'. State top-down policies prioritized economic growth while ignoring the culture and identity of the Adivasis. State agencies manipulate the CNT Act in which land is non-transferable, but in the name of public purposes the land was acquired and given to the capitalist players. Government policy is sometimes contradictory in terms of authorizing this region's access to land and resources.

On the one hand, national-level acts safeguard the Hos' resource access and authority as a tribe of schedule V areas.[12] Moreover, a colonial-era land tenancy act of 1908 renders their lineage-owned and community land not transferable to non-tribal people. But the 'Land Acquisition Act, of 1894' earlier and its recently legislated replacement, the 'Right to Fair Compensation and Transparency in Land Acquisition, Rehabilitation and Resettlement Act, 2013' (LARR Act), enable land to be taken away for mining, prioritizing this as a public purpose despite Hos' constitutional rights. There are many irregularities and contradictions in pluralistic State laws, but these contradictions and irregularities often favoured privileged actors whose economic interest is associated with State building. Mishra and Paty (2010) also talked about the illegality of ACC Company and Tata during land acquisitions at the initial establishment. They mentioned the sub-divisional officers (SDO) of Sadar Chaibasa, who submitted their report to the higher authority of the State. SDO report stated:

> It is clear from the evidence of the facts of the records that some of the raiyats were made to sign or put the thumb impression in the names of dead persons by the company. In some khewats, the name of dead persons and other persons have put false signatures. In some, in place of minors, some others have put the signature. Those people who have their name in the khewat as the ones who have sold their lands are the one who are filing cases against the company, for the restoration of their lands. This shows that the land has been illegally occupied. It seems that through deceit the raiyats have made to sign on the papers and the officers were tricked to get permission for the transfer of the land. In this way Adivasi lands have been illegally alienated by the company.
>
> (Mishra & Paty, 2010, p. 158)

The same thing happened in the case of ACC, a big and old titan of the mining industry, which also acquired land with the help of private contractors and local brokers. Recently, in 2016, people again opposed the lease and area extension of the ACC Cement company. In a public hearing, local villagers said they wanted justice. Here justice is related to the issues of livelihoods, recognition of culture, and participation. Participation in decision-making is one of the crucial dimensions of justice which is never observed in Singhbhum. Any developmental project in the form of extractive industry or infrastructural development of the State does not follow the principle of Free Prior Informed Consent (FPIC). In India, it is replaced by 'consultation', which is also ignored by the Jharkhand government. More particularly, when it comes to issues of development projects and FPIC is to be sorted, the government has to consult with the affected parties but not practise in a very justified manner. It showed how the execution of developmental projects is far from the provisions that deal with self-governance and autonomy in the context of development.

Meanwhile, the process of State building is going on through territorial and institutional expansion and vigorously by top-down development; there remains a pacified and naturalized violence of domination, exploitation, dispossession, and impoverishment. Neoliberal democracy required that they give up their claims as dispossessed, landless subjects and instead compelled them to look to the market to 'deliver' services and jobs. The government would be an 'enabler' of market forces and, rather than redistributing land and wealth, it would enter into 'public-private' partnerships for service delivery. For local economic development, this might mean a new super-maximum prison compound and promises of jobs and skills. In the Noamundi block, mining ruined the local ecology, livelihood, and villagers. Scarcity became an opportunity for intervention. The State, viewed as an external agency, now imposes its populist schemes to reverse the impact and impose a bureaucratic presence.

What the local people have got from the State is subsidized ration on an irregular basis, one electricity pole without electricity and a newly made road. People of this region are suffering from water scarcity. After a long demand of water supply, the company established one water tank some years ago. But during the summer, the villagers of Noamundi block are forced to drink polluted water because the river is dying and also filled with the red dust of iron. One of their morning routines is to stand in a queue to get water for daily routine activities. Earlier, they used to take bath in the river and local tank, and toilet in open space near a river or a *nallah* (natural water tank), but the government order (*Sarkari adesh*) is to make toilets. Block-level government officials forced Panchayati Raj representatives (*mukhiyas*) to construct toilets through MNREGA schemes. Toilet construction was done under MNREGA but again created other water supply problems or toilet availability. Dark, cramped, poorly ventilated, and usually without an arrangement for running water, many toilet constructions were improper and without a roof. The roof was there, but a strong of wind blew it off (personal interview with the beneficiary of the toilet schemes in Gundi Jora, October 2016). One of the central themes in the beneficiary account was the changing life worlds of the Hos of this region and the accommodative politics of the State.

A lady activist who worked with a local non-governmental organization (NGO) in this *hatu* (village) said that Tata Company had made six bio-toilets to facilitate the State cleaning campaign. Still, it is too far from the village. Their everyday tyranny is continued in this form. Getting water, fresh air, and electricity supply is not the only problem in this region; every day after getting the water of their quota, people of this village move to the urban centre in the quest for a daily-wage manual job (Field notes). Indeed, getting work every day is also a routinized task in which villagers rely upon the contractors of mining companies and government schemes. Sundi, a manual labourer of the village of Noamundi basti, said that they get some work through outside contractors but their interest is not the same as the villagers. They pay Rs 200 for 10 hours of work. Their job is to load the minerals into the truck, for which sometimes they are paid Rs 160 (field note). They desire to get as much as possible, but the contractors try to give as little. Even if they get more labourers, the wage comes down to Rs 150 for a day. Getting work is also very hard when the demand for mineral is low in the market.

In Noamundi, where mining covered most of the region, the work and labour are now determined by the contractors, outsiders, and government agencies. The labour wages are not fixed and asymmetrical. It is because many dispossessed Adivasis in the region have lost their traditional livelihood base. The only means of livelihood are government schemes and manual work in the mining site. The multidimensional livelihood base is dying in this region, but State schemes and demand for work MGNREGA are very high. In a household survey, it was found that most of the people have demanded work through MNREGA.

Here, it is evidently clear that government schemes tried to discipline people through the market by increasing their consumption and productivity in terms of monetary value. Access to government services in India is notoriously hard (Harriss & Jeffrey, 2013, pp. 513–515) but through these schemes the State facilitates monetary economics into moral economy of the Adivasi economy. Among the Ho Adivasi groups, those with the slightest knowledge about the government schemes and resource allocation politics capture the politicized right-based services, and they are the exclusive network of uppers and Mundas who are engaged and have regular interaction with block officials. In field areas of Nakahasa, where agriculture is the primary source of livelihood, block officials suggested villagers through village-level officers (panchayat sevak) to get the crop insurance due to continuous drought. Some villagers opted for crop insurance, but very few got insurance claims. State's schemes are not accessible to all because the programme is made to fail (Kumar & Corbridge, 2002; Li, 2005b). There are three reasons for the failure. First, it is expected that the villagers pay the premium in cash. Second, if the Adivasis get trapped in a cycle of debt and default on his agricultural loan – to which his crop insurance scheme is linked – his policy becomes inoperative. Third is that they do not believe in the fantasies of the future gain.

One elder villager of Nakahasa said they did not know about the scheme, and even if they did, how could they pay the premium (Group discussion, October 2016). On the contrary, the Munda of the Nakahasa village was co-opted by the industrialist agents and had an episodic engagement with the block officials to get the insurance claims. Accessing State and market schemes is confined to elites capturing schemes or the villagers who know about the State activities. Therefore, ordinary villagers only have access to at least some benefits of the State in terms of ration, kerosene, and rice. Oppressed villagers, unable to gain more significant benefits, do not consider social mobilization against the government for more substantial benefits from State schemes.

Scott (1976) and Thompson (1971) moral economy of elites oppressing villagers can be useful here to understand how local villagers within the bounds of moral economy and sacral polity do not rebel against the State for this high modernist scheme. In the village of West Singhbhum, the Ho Adivasis also have to tolerate the elites' capture of many schemes and benefit shares, which is inevitable because they avoid the structural mechanism of the State and its bureaucratic complexity (Shah, 2007, 2010).

6.8 Managing the Geography: Development as State Building

This section argues how the noun 'development' is used to manage geography. It has already been discussed how the top-down authoritarian decentralized scheme, policy, and corporate programme try to integrate the Ho. These policies and programmes help the process of the State building by encouraging the Ho Adivasis to participate in it from the below. This participation has been an effective tool for the State making and formation that are directed to 'discipline the Ho adivasi' through schemes. This discipline aims to incorporate the Adivasi polity into the State's idea of governance.

Currently, the State has more than 40 populist State building schemes and welfare programmes that target the Ho community. Most of these have been formulated and implemented under the guidelines of the 20-point programme. However, not all households residing in the Ho village can respond to the various schemes and programmes. Most of the development schemes are related to agriculture, animal husbandry, and allied activities that require assured access to land and fodder in economic terms. Still, many State schemes like MGNREGA, Dhoba construction, pond construction, and toilet scheme are inefficient and inadequate. It is also not popular among the Ho but being implemented through the institutional arrangement of State and coercive mechanisms. After the Shah commission's recommendation against the illegality of the mining companies, 41 companies closed their mining operations. The people absorbed in the neoliberal economy as a fictitious commodity, and a reserve army for sustaining their livelihoods started to migrate from the region. Migration became a living strategy for livelihood, but when drought hit the area, people of the nearby village of Noamundi began to protest against the mining company. They jammed the railway track and did not allow the transportation of iron ore through railway. They demanded work. They needed money to sustain. This only could happen due to the degradation of their lives and ecology. The Member of the Parliament of this district, Laxman Gilua, came to that region and said that he would ask a question in the Lower House of the Parliament about the closure of the mining in iron capital of the Jharkhand. Gilua, who is also the party president of the ruling party of Jharkhand, asked district officials to submit a report on the status of mines. Local leaders and co-opted people argued that if illegalities have been done, State should take the penalty and allow the mining company to function as before.

Only closure of mining was not a problem. The problem was to absorb the dispossessed people into other alternative means. The State never promoted alternative means of production to address the crisis due to capitalism's second contradiction (Connor, 1988). A mining operation of Tata and other big giants started after they gave the penalty and took fresh approval for the mining, but problems of the local people continue till today. People have no work; the surplus people of this region have no choice of job except demanding tedious work from the State and company.

One day in June 2016, villagers blocked the Tata Steel loading point railway track with banners and placards. Local police imposed Section 107 to maintain law and order at the loading point. After the agitation of the local villagers, Member of Parliament Laxman Gilua came to negotiate. Villagers put up their demand to the MP to provide one rack loading manually at the loading point, which would give work opportunities to thousands of villagers, as, currently, they could not arrange two pieces of meal per day. The MP, Gilua, tried to contact the management of Tata Steel, but the management did not respond. However, even though he failed to negotiate with Tata, he succeeded in stopping the agitation for one month. He suggested giving one month to the management to think. After that, they would have to face continuous agitation for an unlimited period until it bowed down to their demands. The MP accused that the management of Tata was stubborn. Nothing happened during those days. Some influential and active villagers participated in the agitation and got some work under government employment schemes. The issue of employment got diluted after some of the villagers got a privileged post in some committees of the village council under JPRA and unconstitutional committees like Gram Vikas Samiti and Adivasi Vikas Samiti. These unconstitutional committees are made neither under the provision of the Panchayati Raj Act nor under the provision of Central PESA. Still, it helps State and local government officials to transform citizens into clients who will be more accountable to the State. In the village of Noamundi, Ho people are cyclically unemployed due to mining operations and even not fully integrated into capitalist production. They can be called 'relative surplus population' (Marx, 1986, pp. 574–606).

Along with the surplus population, the poverty also arises due to uneven capitalist development (Li, 2012). This uneven capitalist development affects the local ecology. It produces proletariats and the formation of labour reserve (Glassman, 2006; Harvey, 2004). The realignment between capital, land, and labour in these areas signals a new form of disconnect.

The realignment among capital, land, and labour in these production regimes signals a new form of disconnect. The purpose is to control offshore land and resources. Offshore labour is also optional. Once dispossessed, opportunities for getting work are highly uneven, as observed in Noamundi region.

In these regions, the scheme to govern through communities, and making them responsible for their fate, has been prominent in the region of Noamundi, especially in the form of micro-credit schemes that require the poor to supply their employment as entrepreneurs (Elyachar, 2005). The State scheme is to gradually produce an effect to develop the theme of community self-reliance, which works as a safety net for the dispossessed Adivasis. It helps them to get reabsorbed into their village economy. This scheme is a technocratic programme to engage the labour within the village and provide an opportunity as one-way tickets 'home' for migrant villagers (Breman & Wiradi, 2002, pp. 2–4).

In mining areas, the State has started some empowerment programmes together with big mining giants like Tata and others. In Noamundi, Jharkhand State Livelihood Promotional Society (JSLPS) hired a private KGVP Rural Enterprises Limited and commissioned a research study to assess the demand and supply in skills and job gaps for youths of this region. Tata and other big giants started an initiative to empower the local tribes in the name of self-help groups and families (*Swarozgaries*) above the poverty line by organizing them into Self Help Groups (SHGs) through capacity building, and provision of income-generating assets. The new groups of SHGs are supposed to go through four stages (Stuart Corbridge et al., 2005). These four stages are: (1) formation, (2) saving, (3), stabilization, and (4) micro-enterprise development. In the first stage, groups are mobilized and motivated by facilitators. Facilitators can be in the form of NGOs or the corporate society of the mining companies; for example, Tata in this region formed a Tata Steel Rural Development Society (TSRDS). The function of this society is to provide ideas and resources to the local villagers that can help them to be self-employed. After formation of SHGs in first stage, the second stage groups begin with savings and intra-group borrowing; in the third stage, SHGs get access to bank loans, and, lastly, in the fourth stage, SHGs are provided with some economic generation activities that have been appropriate for the region under the guideline of the State bureaucracy. The formation of a new institution suits the State's structural adjustment programme and helps the State strengthen its capacity in the name of improvement of the people and region. It enhances the condition for accumulation (Harriss-White, 2003). On the other hand, State also planned to convert some of the villages of Noamundi, Jhikpani, and Manoharpur region into urban centre in the form of 'Nagar panchayat' because people of these villages are disposed from their land and there is no agriculture due to social-ecological transformation of the land. Building houses under the Indira Awas Yojna, creating job opportunities by the appointment of Rojgar Sevak from local youth, implementation of the Forest Right Act 2006 by recognizing the land titles, distribution of Ujjawala Liquid Petroleum Gas Plan, distribution of solar lamps, bicycles and transistors, cloths, mid-day meal during school vacations, infrastructural development by constructing roads and bridges, and construction of buildings through Manki Munda sponsorship of companies are the major plans and soft-developmental interventions of the State.

It is a package of development schemes having new regulative actions and rules. State facilitation and promotion are a kind of optimism and partake in what Chatterjee argues as the 'unscrupulously charitable theoretical gesture of neoliberal ideology' (Chatterjee, 2004, p. 39). State reductionist role is to facilitate modernization thesis observed in field areas. A flow of gifts (transistors, subsidized rice in the name of 'Antyodava Anna Yojana', road, infrastructures, and manual jobs) had a disciplining effect. These disciplining effects can be observed through how the reductionist State deals with the governed subject (Corbridge et al., 2005).

The significant features of the development programme are to integrate the local population into mainstream capitalist economics without caring what the basic needs of the local peoples are. Not only is the soft intervention done, but at the same time land had been leased to 22 mining companies in Saranda region of this district. The corporate nexus with the government started to restructure the productive force of the region. This force of production is within the webs of capitalist relations. These webs of relations mainly act, interact, and react to maintain the dynamic market production process. We extend here the meaning of the term 'production relations' and use it as beyond the economic base. As a matter of culture, knowledge, and ideology with the superstructure, this force of production regularly shapes the Adivasi societal formation. Through this, it is easy to shape the ecological relations of production. The restructuring has been proven a curse for the people whose production and ecological relation are embedded into the ecosystem.

The company did nothing for the region, only looted the resources. Extracting the resources in the name of developing the region is one aspect of the expansion of the commodity frontier. Gradually, as extraction activities expand, it brings the alien apparatus to the tribal land with the State's prior approval. Making a city and urban centre signifies modernity and State making. Still, it also shows how this progress is exclusionary and based on injustice for the Ho Adivasi. The process of dispossession has grievances resulting in the ecology's transformation due to mining like water scarcity, pollution, noise, and dust. It degraded not only the material basis of the land but also people's lives and freedom to pursue and value.

6.9 Development as Structural Violence and Anti-Development

This structural and cultural violence has been perpetuated through symbolic violence by which the oppressed and exploited masses internalize and imitate them (Freire, 2000). The dimension of social transformation in India can be understood through the concept of development as control, development as modernization, or development is for creating a governable object and establishing hegemony. There are similar concepts that also deal with State politics of development to govern the citizen such as 'unequal incorporation and social exclusion' (Nathan & Xaxa, 2012), 'adverse incorporation and social exclusion' (Saha, 1986), 'inclusive and exclusive governmentality' (Bagchi & Chatterjee, 2015), and 'ambivalence of violence'(explained in the following) to understand the State development programme to advance capitalism and State capacity and strength to facilitate market in the name of developing the Adivasi regions. Here, it is essential to understand that development is the development of productive market forces. This type of development structurally reproduced pre-existing society and culture. Strategies of development employed in the Adivasi region aim to alter and restructure societal relations, customary institutions, sacral polity, and alternative modes of production. This process of development is based on structural violence and assimilation. It emerged when the life world of Hos is shaped, in multiple ways, by subordination or repression and the State's will to develop the others. It forced the local polity to interact with the mainstream economics, which misrecognized them as underdeveloped or the community that need State schemes and development for improvement. It engenders 'misrecognition', a kind of 'symbolic violence' (Bourdieu, 1998). All physical violence (material) has a dimension of symbolic (ideal) violence. One can observe these dimensions of violence in Adivasi regions when the Adivasis oppose the developmental industry and the State uses its apparatus to create belief, obedience, and a consciousness to legitimize symbolic violence. To facilitate the capitalist developmental projects, the State uses its monopoly to legitimize physical and symbolic violence. Moreover, dominant discourses might also make people perceive even the most intolerable conditions of existence as acceptable and natural. Such naturalization of unequal power relations and resultant sufferings are routinely carried out through the development process. The development process can be observed in term of political economy of anti-development. This very process of anti-development has been taking place through the developmental policies of the State.

6.10 Conclusion

It is ascertained that the developmental State has two components: structural and ideological. The combination of both ideology and structure has a substantial impact on society because the ideology of development as discourse (Escobar, 2011) is a web of domination. Such a State is essentially one whose mission is to ensure economic development through various means and tactics that can usually be interpreted as development for the improvement of the regions and population. Still, it stimulates high rates of accumulation for achieving economic development. This development process dispossessed many, through schemes and other institutional arrangements, and failed to reabsorb the dispossessed masses within the economy. Such a State established its hegemony through different shadowy practices by recognizing informal and formal institutions to govern a territory and its citizens.

The structural part of the State deals with its capacity to implement and design economic policies and schemes through various arrangements. Through schemes and developmental projects, the State strengthens its ability to negotiate with the Adivasis who have different imaginations and production systems. The State's capacity to inspire the Adivasis is determined through various other agencies at the administration, political, institutional, and customary levels that contribute to State building voluntarily or involuntarily. State presence is felt much in mining areas of Noamundi, whereas in Nakahasa village the State is limited to some uppers and Munda of the village. People of this village depended upon their traditional base and only accessed some populist schemes of the State through the PRI. It is also observed that the State building programmes through schemes and projects incorporated the Ho community into the modern discourse and practice of development.

Notes

1 For more, see (Bauman, 1991, pp. 147–148).
2 A society where co-existence of democratic institutions, capitalist economy, and welfare State can be observed (Marshall, 1950, 1987).
3 The term 'marginalized of the wasteland' is different from the reserve army of labour in Marx since the latter is formed through the process of technological progress under 'capitalist accumulation', while the former belongs to the stage of primitive accumulation (Sanyal, 2007, p. 58). For brief discussion on this issue, see also, http://www.southasia.ox.ac.uk/sites/sias/files/documents/ali%20jan%20sanyal-review-final.pdf.
4 The root of establishing the panchayat can be traced from the Bihar Panchayat Raj Act, 1948; Gram panchayats had been established on a statutory basis in Chotanagpur and Santhal Parganas (Troisi 1978: 152).
5 For more information about the PESA, see also, http://pesadarpan.gov.in/documents/26993/0/PESA+Act+1996.pdf/7a0e6cf5-3d28-4fce-a15e-39e513961cc8.
6 Personal interview (March 2016).
7 For more discussion on the controversy and filed court case, see https://indiankanoon.org/doc/18838/.
8 Since the colonial times, Manki Munda has been collecting the revenue from the village and depositing it to the State administration. This practice is continuing but within State laps.
9 For more detail on JPRA, see also, http://www.jharkhandpanchayats.gov.in/documents/6183151/0/1345192397587_Jharkhand_Panchayat_Raj_Act_2001_English3186.pdf.
10 There are many cases of resource conflict observed during field study. Some of the important cases protested by the villagers are: ACC extension of lease and acquisition in Hat Gamharia, Tonto, and Jhikpani for limestone mining; quartz mining in Abaru village of Khuntpani block; and stone crusher establishment in Noamundi block. In all cases of resource conflict, villagers through Gram Sabha gave a petition to district authority against the company mining.

11 Reddy and Haragopal (1985) used terms such as 'fixers', 'brokers', and *pyraveekar* to refer to those who thrived in spaces that needed manoeuvring precisely because of the nature of a peculiar type of State–society relationship, i.e., structural violence (Gupta, 2012). Scholars like Mitra (2002) called them village leaders (*Gaon ka neta*). They are engaged in realms of local politics, bureaucracy, and collective protests.

12 For more discussion on Schedule V of Indian constitution, see, https://www.mea.gov.in/Images/pdf1/S5.pdf.

Chapter 7

Conclusion

The study is concerned with the State's 'will to develop Adivasi'. It is interrogated through the lens of political ecology. It deals with how the State views developmental projects and schemes, as an immanent process and activity of capitalist development that interacts with the Adivasi ecology and economy, taken broadly in reference to natural resources, sacral polity, alternative imagination of Adivasi and extractive industries, backed up ideologically and economically by the State agencies through several discourses, schemes, and institutions. It explores the understanding of State formation through developmental intervention in predominately resource-rich region of the Hos in Jharkhand. It addresses the interrelated issues of development-induced dispossession, resistance, ecological transformation, and Adivasi resistance against the State extractive projects, shadowy practices of democracy, illegalities, and State building.

This study set out to analyse how State stimulates the extractive players in resource-rich region of the Ho that induces power asymmetry and trajectories of injustice and assimilation. It has been done through an extended case study research methods and participatory observation on the interaction between the representative of the capitalist State and the defender of the moral economy in West Singhbhum by looking at the resource access struggles and resistance connected with the issues related to resistance, changes, justice, and assimilation.

The research was approached from the perspective of political ecology by using participants' observation and ethnography. This study has also used concepts and insights from the other disciplinary fields to enrich the potential of the political ecology. In particular, I have used the concept from developmental sociology, neo-Marxism, peasant studies, environmental and critical sociology and economic anthropology, as well as the theories of resource governance, State formation, and building.

7.1 Conclusion from the Previous Chapters

Specific chapters of this study (Chapters 3–6) address the four research objectives on the basis of primary and secondary evidences collected from the field

DOI: 10.4324/9781032637990-7

and pre-existing literature, respectively. Most of the collected cases and narrations deal with the accumulation, land dispossession and expansion of the extractive industries. In Subsections 7.2.1–7.2.4, I present the conclusions drawn from the chapters that deal with the particular research questions.

7.1.1 Various Debates on State and Development through the Political Ecology Lens

Chapter 3 deals with the review of the development and its discontent. It talks about how State developmenalism is merely a neoliberal project and capitalist constructed discourse. It is not only about economics, but it also ensures to promote political and ideological conditions for capitalists to tackle the other economic and political institutions in the name of national integration and growth. Associated with the ideology of market fetishism, this capitalist form of State-mediated development only promotes the policy of capital. It deals with how State's (colonial to post-colonial) development policy for the Adivasis was a means of the resource control of the region that is even continued till date through developmental discourses and projects that have been protested and resisted by the Adivasis of this region.

Firstly, development as a discourse and project is a kind of governmentality that defines development on the basis of a projected desirable society. Thus, development as a vision shows that as a process it takes account of the constant growth and desirable improvement, modernization, control, and resource exploitation, which also facilitated migration, dispossession, and marginalization. It is mainly concerned with the capital-centric growth and linear progress model. The colonial State also sought to civilize the wild and unruly Hos through their colonial modernization scheme that enhanced their imperial agenda. The colonial State pursued a policy of non-interference in the working of the Hos' village at the local level and at the same time enhanced the land revenue system and forest policy to generate more surplus to British Raj.

Secondly, the review has shown that the post-colonial idea of the State in development, popularly in the vernacular context known as *vikas* (Development), refers to being accepted by majority of Indians as a legitimate activity promoted by the State (Rangan 1997). The post-colonial model of development in India is based on two interrelated processes: one is related to the immediate use of natural resources and the other is related to transforming the people and communities against their will into a hegemonized, dispossessed class.

Thirdly this chapter demonstrates that development often entails 'accumulation by dispossession' (Harvey, 2004), and this accumulative project is a kind of high modernist development scheme of the State. After the formation of Jharkhand, the State's continuing promotion of the establishment of extractive factories to extract natural resources surround the commodity boom, but it also transforms this harmonious place into a conflict zone. The class-based struggle over the control and access of natural resources and distribution of rents is predominant in these areas, which also give a description of the way the State engages, negotiates, and manages the ecology to promote commodity boom. As mentioned in the earlier discussion, looking at the resistance and protest against the State is a vital element for understanding the State development projects.

Finally, this chapter shows that development as discourses, projects, and processes is embedded within the series of changing contexts, processes, and forces that transformed the region.

7.1.2 State Formation Process and Change

Chapter 4 has provided a brief account of the historical, cultural, and political background of the processes of change in the Hos-predominant region of West Singhbhum. It explored how development as a process and activity fuelled through State formation has shaped the local ecology of the region. It addressed the second research objective. It explained the genealogy of State formation and changes among the Ho Adivasis of West Singhbhum through the five theses of political ecology.

Firstly, the top-down policies and projects currently boosted by neoliberal policies and State's 'will to improve the region' through industrialization fail to fulfil the aspiration of the local inhabitant residing in the villages. The influence of these policies aimed at promoting economic activities based on the extractive of natural resources induces exclusion, resistance, and marginalization, and produces a new identity and governance system.

It shows that territorialization, conservation, and commodification brought marginalization, which is the sum of unequal social situations that reduce options for life (Blaikie and Brookfield, 1987; Li, 2014; Robbins, 2012). The chapter talked about the five theses of political ecology to deal with the issues of changes of West Singhbhum. State's development narrative argues that resource extractive industries is for the betterment of the local people – the official account claims that the production of minerals and life choices of the Adivasis has been increasing but paradoxically Ho Adivasis maintained that State projects have transformed their ecology, culture, and livelihood. It can be said that the State exercises its power and control through developmental projects but never guaranteed beneficial outcomes and its promised delivery. The Adivasi region of Jharkhand has a history of continuously failed improvement schemes and diagnoses. State-planned extractive industry failed to deliver their promises and the diagnosis has been also incomplete but it routinely produces effects in the form of marginalization from the colonial times onwards.

Secondly, the theses of political ecology showed that State formation is hafted to ecological degradation, territorialization projects, and dispossession, which also produced resistance and different kind of political identities.

7.1.3 Developmental Projects have Aggravated Conflict and Resistance Over Natural Resources

The discussion in Chapter 5 leads to the following conclusions. It has aimed to understand what resources meant for Ho. Why have they been struggling and resisting against the mining companies and developmental plans of the State to save their resources? It also elucidates how the extractive developmental project and its illegalities are based on power asymmetry. On the basis of arguments put forth in the previous chapters regarding the State formation and changes in terms of dispossession and marginalization, this chapter tried to uncover the State character and its illegalities. The cases analyzed in Chapter 5 lead to the following conclusions.

Firstly, government support in the form of incentives and policies to promote extractive activities induces resistance that also stimulated unequal access to resources. Government support to extractive activities and forms of access to and control over resources is a part of the 'political economies of power and accumulation' underlying socio-environmental conflicts' (Peluso, 2009, p. 40).

Secondly, the State's myopic encouragement to companies through policies and plans showed that the role of the State as anti-Adivasi (material, meaning, and capabilities) valuation attached by losers and winners is found to be different. For protestors, it was their territory which gave them an identity and a sense of belongingness, whereas for the others, mainly industrialists and State agencies, mines are a profitable commodity.

Thirdly, the government's support to extractive industries and activities has benefitted the corporate houses, without considering the Adivasis who have multiple relationships with the local ecology. The State's heedless encouragement to companies through policies and plans showed 'how State is doing a discriminating intervention' (Scott, 1998a, p. 3).

Shared culture plays a vital role in resistance, for instance, *Sasandari* (stone slob as a symbol of social solidarity and self-governance) and *Jahira* (sacred spaces) of the Hos are pivotal for the ecosystem and landscape. Any encroachment in their sacred space induces resistance.

Cases discussed in Chapter 5 lead to following conclusions about how the Hos have used everyday resistance to oppose the State extractive developmental projects. Sometime their resistance also came in the form of writ petitions, slogan marches, and negotiations. The cases and discussion showed that the threat of livelihoods and the resulting dispossession and exclusion from their own land and life choice are the main reasons for resistance. The manipulation of the CNT Act and constitutional safeguard provisions are also a reason for the everyday resistance.

Fourthly, some of the civil society and activists who worked in these areas regularly use different mechanisms and strategies with different capitalist players. Their slogan '*humara goan me humara raj* (our village our rule) and

Abub Raj (self-rule) critiqued the top-down model of State development projects.

The discussion in Chapter 5 also confirms the existing political ecology literature. It showed that there are many reasons for the resistance, and all the factors are deeply embedded in each other. Political ecology observed conflict through the concept of accumulation through dispossession (Harvey, 2004). In this vein, some of the other concepts, that is, power asymmetry (Bryant & Bailey, 1997), talk about the actor's desire to control or access the nature that induces conflict and assimilation. Escobar's (2006b) concept of assimilation is also important because in many places where mining is ongoing or failed to be operationalized, people have opposed the State's move to incorporate them into the modern discourse, processes, and practices of development. It also induces cultural conflict and resistance because they think that Singhbhum (Kolhan) is meant for Hos, and that, if once displaced or dispossessed, they may have to lose their cultural identity.

7.1.4 *Negotiating* Change: *Customary Practices and State Building Programme*

Chapter 6 has demonstrated that in resource-conflict areas, the top-down developmental schemes also help the State to strengthen its capacity. State building is about controlling violence, establishing legitimacy, and building capable and responsive institutions so as to foster a shared sense of the public realm. It showed that assimilation and co-optation may be manifested in multiple and varied practices through the implementation and monitoring of schemes. At the local level, it also produced political fixers, brokers, and middlemen.

Firstly, government employment schemes are very relevant and act as mediating actors. The schemes' implementation through different line agencies and other institutions showed the structural and relational mechanism the access that facilitates the assimilation process. It is used to absorb the dispossessed people within the village territory.

Secondly, Decentralization of resource governance was not practised at the ground level. The Kolhan administration (customary system) is degrading day by day, and State bureaucratic power tackles its power in the name of developing the region. Governance is a State agenda but it is based on the secular kind of alternative government in which Adivasi cultural beliefs and religion have no space (Shah, 2010). It is promoted within the framework of a post-colonial capitalist State where the economy determines the validity of the institution

Thirdly, populist politics of the State and the State's will to improve the Adivasis through several schemes were instrumentally deployed for electoral gain but it is translated as a participatory democracy. The targeted programme is not based on general welfarism, but it is the subject of a developmental State. The Adivasi local chiefdoms were cast as clients of the bureaucratic government. The State asked them to make a plan for the village but the sketches, directions, and blueprints of plans and policies are already determined at the top level. Panchayati Raj Institutions (PRIs) became cast as a client of the bureaucratic government through the extension of the targeted policies such as Dhoba, toilet scheme, Pradhan Mantri Awas Yojna etc., which are implemented and monitored by the political party committees. They vastly extended the patronage politics, and through this the neoliberal State reached down to the local level and extended political strength.

Fourthly, at the time of conflict, villagers through traditional institution under the leadership of Munda challenged the State's actions. This is the way they resisted the State and legitimated their access. At the time of resource conflict, it was observed that (mining-induced ecological degradation, and protests and resistance against land acquisition) villagers through Gram Sabha under the presidentship of Munda gave a written petition to the deputy collector against the mining operation. If the company succeeded in co-opting the local leader including Munda, then the resistance against the company or extractive industry came into the negotiable stage. The process of co-optation also induced assimilation of the traditional institution and their functions, whereas in the case of government schemes villagers looked at the PRIs. Central Panchayats Extension to Scheduled Areas Act (PESA) is not exercised in this region also favoured corporate to acquire the land without the prior permission of Gram Sabha.

Fifthly, the illegalities were observed in the field area where State has established many informal institutions that violate the constitutional spirit and legal provisions. This also delimited the power of traditional chiefdoms. The dimension of environmental justice (distribution, recognition, capabilities, and participation) are undermined at the time of implementation of schemes and projects.

This study further enhanced the political ecology knowledge on State building in conflict-prone resource-rich region of the Adivasis. It has demonstrated that the following:

1) Co-optation and assimilation weaken the everyday resistance and strengthen the State capacities.
2) Accumulation by dispossession is backed up through government intentional schemes like the Mahatma Gandhi National Rural Employment Guarantee Act (MNREGA) and other livelihood schemes. It acts as a mediating actor. It is used to absorb the dispossession of the area.
3) Access to resources and control over resources are influenced by the State policies and acts that can be observed in the form of multiple governmentality.
4) State projects recognized Adivasi institutions for political purposes and for co-optation to lessen the resistance.

5) Shadowy democratic politics and practices of the State succeed in legitimating the State actions through the different institutions including the customary institutions.

The discussion about the cases of the expansion of the extractive industry and the State formation among the Hos cannot be generalized fully because this region has a peculiar governance system (traditional and modern) and has an overlapping legislature and tenure system, but this study can give a better understanding of the resistance against the State through the analysis of the strategies and the politics used by the resisting actors and the conditions that facilitate resistance and everyday negotiation. It also helps to understand the process and mechanism of the State government, which adopted shadowy practices to counteract the negative impact of the resource extractive industry in the name of participatory democracy and governance.

7.1.5 Contribution to Knowledge and Practices

The findings of the study open a new vista for further research in the field of the political ecology of the State and the Adivasis in terms of contradiction and compromise relationships. True to the principle of the ethnography of the State and political ecology, this study contributed a better understanding of the State formation among the Ho Adivasi habitat, where State-stimulated extractive projects in the form of development as project and activity have accelerated conflicts, everyday resistances, and changes. It also reconstructs the knowledge and practices of State through the practices of sociology of emergence and sociology of absences by giving voices to the Hos.

7.2 Contributions to Knowledge to Understand the Trialectics

This study is based on political ecology scholarship and has contributed an enhanced understanding of the developmental State in the mineral resource-rich region of Ho Adivasis. Some of the major contributions of the study are as follows:

1) It provides a nuanced understanding of the role of the State in the acceleration of extractive industries that shaped the ecology and economy of the region by unequal distribution.
2) This study also gives special attention to the legalities and illegalities done by the State to facilitate extractivism by undermining the issues of justice and plurality of the relation Hos has with their territory.
3) It also dealt with the power relation and institutional politics and different meanings of the development that determine the trialectics (State, society, and rest of nature, or the interaction between politicized interaction between socio-nature and State).

7.3 Policy Relevance

The policy relevance of academic research findings has remained elusive, despite efforts to align them with the needs of the state and society. Certain academic activists, to reach out to people with the right perspective, initiate debates with the State administration involving line agencies, bureaucracies, and organizations in charge of planning and implementing policy and often find them irrelevant. They have little power over how their work is or is not used. Whether or not such efforts will result in meaningful policy reform depends upon where power lies and how it is being used (Blaikie, 1985; Neumann, 2008).

Following their statement, it can be stated that offering policy recommendations assume lesser importance wherein the State defines its priorities, agendas, and politics of development through various discourses ranging from preservation and conservation to acquisition, displacement, rehabilitation, and reconstruction. The writings of a researcher must trigger the political will of the public representative. The Gandhian concept of Gram Swaraj has its relevance even today if the State honours the self-rule of the tribal people and the victims feel as victors.

The recommendation is based on the discussion presented in this study about the manipulation of laws and facilitation of mining companies by giving more priority to the economy over ecology, which has resulted in resistance and has created socio-environmental changes in the entire geographical region which had hitherto remained virgin and affected the livelihoods and primordial institutions of the Ho Adivasis. Consultation with local populations should be a best practice if the State intends to act as a facilitating institution in favour of the participatory democracy in its spirit. The PESA, the Chotanagpur Tenancy Act (CNTA), and other acts should be implemented in this region with the true spirit of the democracy and constitution to redress the historical injustice made to the Adivasis.

7.4 Area for Future Research

A few areas identified for future research are given in the following text.

More empirical and longitudinal research on political ecology would help to understand the dynamics of the shadowy practices in a democratic State. Further, this research can be extended to deconstruct the internal politics of subordination of the State, the domineering position of the market economy, and the rise of capitalism. The customary institutions at the grassroots and their resistance in response to the State-sponsored development programmes as a prelude to the modern State formation can be extended in greater detail. The resistance movement of the Adivasis during the post-liberal economic era juxtaposing pragmatic political ecology and the private wisdom of the capitalistic order can also be an upcoming area of research. Based on secondary sources of data, from a political ecological perspective, a comparative study can be carried out between the pristine primitive State formation and the modern State formation. Another dimension of the study can also be targeted on the history of counter-hegemonic movements against the oppressive role of the State and other apparatus like the judiciary, executive, and success and failure of the current sporadic movement of outfits.

References

Abrams, P. (1988). Notes on the difficulty of studying the state. *Journal of Historical Sociology*, *1*(1), 58–89. https://doi.org/10.1111/j.1467-6443.1988.tb00004.x

Abu-Lughod, L. (1990). The romance of resistance: Tracing transformations of power through Bedouin women. *American Ethnologist*, *17*(1), 41–55. https://doi.org/10.1525/ae.1990.17.1.02a00030

Adams, J. (2001). Culture and economic development in south Asia. *The American Academy of Political and Social Science (AAPSS)*, *573*, 152–176.

Agnew, J. A. (2013). Territory, politics, governance. *Territory, Politics, Governance*, *1*(1), 1–4.

Agrawal, A. (2001). Collective action, property rights and decentralisation in resource use in India and Nepal. *Politics and Society*, *29*(4), 483–613.

Agrawal, A. (2005). *Environmentality: Technologies of government and the making of subjects*. Durham: Duke University Press.

Althusser, L. (1971). *Lenin and philosophy and other essays*. London: New Left Books.

Amin, S. (1974). Modes of production and social formations. *Ufahamu: A Journal of African Studies*, *4*(3), 57–97. Retrieved from http://escholarship.org/uc/item/4rh67483%5CnAcknowledgements

Amin, S. (1976). *Unequal development*. New York: Monthly Review Press.

Amsden, A. (2000). Comment: Good bye dependency theory, hello dependency theory. *Studies in Comparative International Developemnt*, *38*(1), 32–38.

Anten, L. (2009). *Strengthening governance in post-conflict fragile states*. Clingendael' Conflict Research Unit. The Hague: Netherlands Institute of International Relations.

Apffel-Marglin, F., & Marglin, S. (Eds.). (1990). *Dominating knowledge, development, culture and resistance*. London: Oxford University Press.

Appadurai, A. (1988). *The Social life of things: Commodities in cultural perspective*. New York: Cambridge University Press.

Areeparampil, M. (1993). Socio-religious and cultural movements among the Ho tribals of Singhbhum district. In M. Miri (Ed.), In *Continuity and change in tribal society* (pp. 396–437). Shimla: Indian Institute of Advanced Studies.

Areeparampil, M. (1996). Displacement due to mining in Jharkhand. *Economic and Political Weekly*, *31*(24), 1524–1528.

Areeparampil, M. (2002). *Struggle for swaraj: A history of Adivasi movements in Jharkhand (from the earliest times to the present day)*. Chaibasa: Tribal Research and Training Centre.

Ashokan, A. T., & Rangarajan, G. (2011). Writing the subaltern into being: Tribals and nature in Mahasweta Devi's Petrpdatyl. *Contemporary Research in India*, *1*(2), 4–10.

Aunger, R. (2004). *Reflexive ethnographic science*. Oxford: Alta Mitra Press.

Bagchi, A. K. (1982). *The political economy of underdevelopment*. Cambridge: Cambridge University Press.

Bagchi, A. K., & Chatterjee, A. (2015). *Marxism with and beyond Marx*. New York: Routledge. https://doi.org/10.1017/CBO9781107415324.004

Baginski, O. S., & Blaikie, P. M. (2013). *Forests, people and power: The political ecology of reform in south Asia*. Earthscan. Retrieved from https://books.google.com/books?id=flzipucSBLYC&pgis=1

Ball, V. (1880). *Jungle life in India or the journeys and journals of an Indian geologist*. London: De La Rue & Co.

Ballvé, T. (2012). Everyday state formation: Territory, decentralization, and the Narco landgrab in Colombia. *Environment and Planning D: Society and Space*, *30*(4), 603–623.

Baran, P. A. (1957). *The Political economy of growth*. New York: Monthly Review Press.

Bardhan, A. B. (1973). *The tribal problem In India*. New Delhi: Communist Party Publication.

Bardhan, P. (1984). *The political economy of development in India*. New Delhi: Oxford University Press.

Bardhan, P. (2011). Industrialization and the land acquisition conundrum. *Development Outreach*, *13*(1), 54–57.

Bassett, T. J. (1988). The Political ecology of peasant-herder conflicts in the northern Ivory Coast. *Annals of the Association of American Geographers*, *78*(3), 453–472. https://doi.org/10.1111/j.1467-8306.1988.tb00218.x

Bauman, Z. (1991). *Modernity and the holocaust*. Ithaca, New York: Cornell University Press.

Baviskar, A. (1997a). Ecology and development in India: A field and its future. *Sociological Bulletin*, *46*(2), 193–207.

Baviskar, A. (1997b). *In the belly of the river: Tribal conflicts over development in the Narmada Valley*. New Delhi: Oxford University Press.

Baviskar, A. (2003). Between violence and desire: Space, power, and identity in the making of metropolitan Delhi. *International Social Science Journal*, *55*(175), 89–98. https://doi.org/10.1111/1468-2451.5501009

Bebbington, A. (1999). Capitals and capabilities: A framework for analyzing peasant viability, rural livelihoods and poverty. *World Development*, *27*(12), 2021–2044.

Bebbington, A. (2000). Recountering development: Livelihood transitions and place transformations in the andes. *Annals of the Association of American Geographers*, *90*(3), 495–520.

Bebbington, A. (2010). Extractive industries and stunned state: Conflict responsibility and institutional change in the Andes. In K. Ravi Raman & R. D. Lipschutz (Eds.), *Corporate social responsibility: Comparative critiques* (pp. 97–116). London: Palgrave Macmillan.

Bebbington, A. (2011). An Andean avatar: Post-neoliberal and neoliberal strategies for securing the unobtainable. *New Political Economy*, *16*(1), 131–145.

Bebbington, A. (2012). Underground political ecologies: The second annual lecture of the cultural and political ecology specialty group of the association of American geographers. *Geoforum*, *43*(6), 1152–1162. https://doi.org/10.1016/j.geoforum.2012.05.011
Bebbington, A. (2013). *Natural resource extraction and the possibilities of inclusive development: Politics across space and time. (No. 21). ESID, Identifying routes to social justice.* ESID Working Paper No. 21. Manchestor. Retrieved from https://ssrn.com/abstract=2386711 or http://dx.doi.org/10.2139/ssrn.2386711
Beck, U. (1992). *Risk society: Towards a new modernity*. London: Sage.
Beck, U. (2012, July 11–15). The social and political dynamics of the world at risk: The cosmopolitan challenge. In *26th annual congress of the Association of European Schools of Planning*. Ankara.
Becker, M. (2012). Building a plurinational Ecuador: Complications and contradictions. *Socialism and Democracy*, *26*(3), 72–92.
Bergh, S. (2004). Democratic decentralisation and local participation: A review of recent research. *Development in Practice*, *14*(6), 780–790.
Berry, S. (1988). Concentration without privatization? Some consequences of changing patterns of rural land control in Africa. In S. Reyna & R. Downs (Eds.), *Land and society in contemporary Africa* (pp. 53–75). Hanover, NH: University Press of New England.
Bhaduri, A. (2016). *Malignant growth*. New Delhi: Aakar Publication.
Biersack, A., & Greenberg, J. B. (2006). *Reimagining political ecology*. Durham: Duke University Press.
Blaikie, P. (1985). *The political economy of soil erosion in developing countries*. NewYork: Longman Development Studies.
Blaikie, P. M., & Brookfield, H. (1987). *Land degradation and society*. New York: Routledge.
Blaser, Mario, Feit, Harvey A., & McRae, Glenn (Ed.). (2004). In the way of development: Indigenous peoples, life projects and globalization, 362. Retrieved from https://books.google.com/books?id=0UwtqpeBtvQC&pgis=1
Borras, S. M., Franco, J. C., Gomez, S., Kay, C., & Spoor, M. (2012). Land grabbing in Latin America and the Caribbean. *Journal of Peasant Studies*, *39*(3), 845–872.
Bourdieu, P. (1998). *Acts of resistance: Against the new myths of our time*. London: Cambridge Polity Press.
Bradley-Birt, F. B. (1874). *Chota Nagpur: A little-known province of the empire*. London: J. Murry.
Brecher, J., & Costello, T. (1994). *Global village or global pillage: Economic restructuring from the bottom up*. Boston: South End Press.
Breman, J., & Wiradi, G. (2002). *Good times and bad times in rural Java*. Leiden: KITLV Press.
Brinkerhoff, D. W. (2005). Rebuilding governance in failed states and post-conflict societies: Core concepts and cross-cutting themes. *Public Administration and Development*, *25*(1), 3–14.
Brock, A., & Dunlap, A. (2018). Normalising corporate counterinsurgency: Engineering consent, managing resistance and greening destruction around the Hambach coal mine and beyond. *Political Geography*, *62*, 33–47.
Brookfield, H. C. (1975). *Interdependent development*. London: Methuen.

Bryan, J. (2012). Rethinking territory: Social justice and neoliberalism in Latin America's territorial turn. *Geography Compass*, *6*(4), 215–216.

Bryant, R. L. (1992). Political ecology An emerging research agenda in Third-World studies. *Political Geography*, *11*(1), 12–36.

Bryant, R. L. (1998). Power, knowledge and political ecology in the third world: A review. *Progress in Physical Geography*, *22*(1), 79–94. https://doi.org/10.1177/030913339802200104

Bryant, R. L. (2004). Putting politics first: The political ecology of sustainable development. *Global Ecology and Biogeography Letters*, *1*(6), 164–166.

Bryant, R. L., & Bailey, S. (1997). *Third world political ecology. Geography* (Vol. 1). London: Routledge. Retrieved from http://books.google.com/books?id=oVfEAfVBYVoC&pgis=1

Bryman, A. (2004). *Social research methods*. London: Oxford University Press.

Bryman, A., & Burgess, R. G. (1994). Developments in qualitative data analysis: An introduction. In R. G. Bryman & Alan Burgess (Ed.), *Analyzing qualitative data*. (1st ed., pp. 217–220). London and New York: Routledge. https://doi.org/10.4324/9780203413081

Burawoy, M. (1998). The Extended case method. *Sociological Theory*, *16*(1), 4–33. https://doi.org/10.1111/0735-2751.00040

Burawoy, M. (2004). Public sociologies: Contradictions, dilemmas, and possibilities. *Social Forces*, *82*(4), 1603–1618. https://doi.org/10.1353/sof.2004.0064

Bury, J. (2005). Mining mountains: Neoliberalism, land tenure, livelihoods, and the new Peruvian mining industry in Cajamarca. *Environment and Planning A*, *37*(2), 221–239. https://doi.org/10.1068/a371

Buscher, B. (2010). Anti politics as political strategy: Neoliberalism and transfrontier conservation in southern Africa. *Development and Change*, *41*(1), 29–51.

Castree, N. (1995). The nature of produced nature: Materiality and knowledge construction in marxism. *Antipode*, *27*(1), 12–48.

Castree, N. (2008). Neoliberalising nature: Processes, effects, and evaluations. *Environment and Planning A*, *40*(1), 153–173. https://doi.org/10.1068/a39100

Cavanagh, C. J., & Benjaminsen, T. A. (2015). Guerrilla agriculture? A biopolitical guide to illicit cultivation within an IUCN Category II protected area. *Journal of Peasant Studies*, *42*(3–4), 725–745.

Cederlöf, G., & DasGupta, S. (2016). *Subjects, citizens and law: Colonial and independent India*. London: Routledge.

Cernea, M. M. (1988). *Involuntary resettlement in development projects: Policy guidelines in World Bank-financed projects*. World Bank Publications.

Cernea, M. M. (1990). Internal refugee flows and development-induced population displacement. *Journal of Refugee Studies*, *3*(4), 320–339.

Chakrabarti, A., & Dhar, A. (2009). *Dislocation and resettlement in development*. London: Routledge.

Chakravorty, S. (2013). *The price of land acquisition, conflict, consequence*. London: Oxford University Press.

Chambers, R. (2006). *Poverty Unperceived: Traps, biases and agenda (No. 270). IDS Working Paper*. Sussex.

Chandra, U. (2013a). Beyond subalternity: Land, community, and the state in contemporary Jharkhand. *Contemporary South Asia*, *21*(1), 52–61.

Chandra, U. (2013b). Liberalism and Its other: The Politics of primitivism in colonial and postcolonial Indian Law. *Law & Society Review*, *47*(1), 135–158.

Chandra, U. (2015a). Rethinking subaltern resistance. *Journal of Contemporary Asia*, *454*(July), 563–573. https://doi.org/10.1080/00472336.2015.1048415

Chandra, U. (2015b). Routledge handbook of contemporary India. In K. A. Jacobsen (Ed.), *Routledge handbook of contemporary India* (pp. 297–309). London: Routledge.

Chatterjee, A., & Das, T. C. (1927). *The Hos of Seraikela* (No. paper 1). Calcutta.

Chatterjee, P. (1998). Development planning and the Indian state. In T. J. Byres (Ed.), *State, development planning and liberalisation in India* (pp. 82–103). Oxford: Oxford University Press.

Chatterjee, P. (2004). *The politics of the governed: Reflections on popular politics in most of the world*. New York: Columbia University Press. https://doi.org/10.1057/palgrave.cpt.9300318

Chibber, V. (2003). *Locked in place: State-building and late industrialization in India*. Princeton, NJ: Princeton University Press.

Chomsky, N., Meyer, L., & Maldonado, B. (2010). *New world of Indigenous resistance: Noam Chomsky and voices from North, South, and Central America*. San Francisco: City Lights Publishers.

Cloke, P., Crang, P., & Goodwin, M. (2013). *Introducing human geographies* (3rd Ed.). London: Routledge.

Cohn, B. (1996). *Colonialism and its forms of knowledge*. Princeton: Princeton University Press.

Cohn, B. S. (1961). From Indian status to British contract. *The Journal of Economic History*, *21*(4), 613–628.

Collier, P. (2010). The Political economy of natural resources. *Social Research: An International Quarterly*, *77*(4), 1105–1132. Retrieved from http://socialresearch.metapress.com/index/R7577J14T18112G5.pdf

Connor, J. O. (1988). Capitalism, nature, socialism: A theoretical introduction. *Taylor & Francis*, *1*(1), 11–38. Retrieved from http://www.tandfonline.com/doi/abs/10.1080/10455758809358356#.VbXDZLOqqko

Corbridge, S. (1982). Interdependent development? Problems of aggregation and implementation in the Brandt Report. *Applied Geography*, 2, 253–265.

Corbridge, S. (1988). The ideology of tribal economy and society: Politics in the Jharkhand, 1950–1980. *Modern Asian Studies*, *22*(1), 1–42.

Corbridge, S. (1990). Post-Marxism and development studies: Beyond the impasse. *World Development*, *18*(5), 623–639. https://doi.org/10.1016/0305-750X(90)90014-O

Corbridge, S. (1993a). Marxisms, modernities, and moralitiee: Development praxis and the claims of distant strangers. *Environment and Planning: Society and Space*, *11*(1976), 449–472. https://doi.org/10.1068/d110449

Corbridge, S. (1993b). Ousting singbonga: The struggle for India's Jharkhand. In P. Robb (Ed.), *Dalit movements and the meanings of labour in India* (pp. 121–150). London: Oxford University Press.

Corbridge, S. (2009). The political economy of development in India since independence. In *Routledge handbook of south Asian politics: India, Pakistan, Bangladesh, Sri Lanka and Nepal*. Retrieved from http://eprints.lse.ac.uk/20381/1/The_political_economy_of_development_in_India_since_independence_(author_final).pdf

Corbridge, S., Williams, G., Srivastava, M., Veron, R., & Véron, R. (2005). *Seeing the state: Governance and governmentality in India*. Cambridge: Cambridge University Press. Retrieved from http://eprints.lse.ac.uk/2880/

Cowen, M., & Shenton, R. W. (1996). *Doctrines of development*. London: Routledge.

Crewe, E., & Axelby, R. (2013). *Anthropology and development: Culture, morality and politics in a globalised world*. Cambridge: Cambridge University Press.

Cuveller, J., Vlassenroot, K., & Olin, N. (2014). Resources, conflict and governance: A critical review. *Extractive Industries and Society*, *1*(2), 340–345.

Dalby, S. (2003). Environmental Insecurities: Geopolitics, resources and conflict. *Economic and Political Weekly*, *38*(48), 5073–5079.

Dalton, E. T. (1872). *Descriptive ethnology of Bengal*. Calcutta: Office of the Superintendent of Government Printing.

Damodaran, V. (1997). Environment, ethnicity and history in Chotanagpur, India, 1850–1970. *Environment and History*, *3*(3), 273–298. https://doi.org/10.3197/096734097779555854

Damodaran, V. (2006). Colonial constructions of the "tribe" in India: The case of Chotangapur. *Indian Historical Review*, *33*(1), 44–73. https://doi.org/10.1177/037698360603300104

Dandekar, A., & Chitrangada, C. (2011). *PESA, LWE and governance: Concerns and challenges in India's tribal districts*. Anand: Institute of Rural Management.

DasGupta, S. (2009). Accessing nature: Agrarian change, forest laws and their impact on an adivasi economy in colonial India. *Conservation and Society*, *7*(4), 227. https://doi.org/10.4103/0972-4923.65170

DasGupta, S. (2011). *Adivasis and the raj: Socio-economic transformation of the Hos*. Hyderabad: Orient Blackswan.

Dayal, K. (2010). Silence that is disposable. *Combat Law*, 1–3.

De Soysa, I., & Neumayer, E. (2007). Resource wealth and the risk of civil war onset: Results from a new dataset of natural resource rents, 1970—1999. *Conflict Management and Peace Science*, *24*(3), 201–218.

Debord, G. (1994). *The society of spectacle*. New York: Zone Book.

Deibert, R. J. (1999). Harold Innis and the empire of speed. *Review of International Studies*, *25*(2), 273–289.

Dell'Angelo, J., D'Odorico, P., Rulli, M. C., & Marchand, P. (2017). The tragedy of the grabbed commons: Coercion and dispossession in the global land rush. *World Development*, *92*, 1–12. https://doi.org/10.1016/J.WORLDDEV.2016.11.005

Denny, J. (2008). *The spirit world of the Hos*. Chaibasa: Xavier Ho Publications.

Devalle, S. B. C. (1992). *Discourses of ethnicity: Culture and protest in Jharkhand*. Delhi: SAGE Publications.

Devi, M. (1981). Witch-sabbath at Singhbhum. *Economic and Political Weekly*, *16*(40), 1595–1597.

Dhebar, U. N. (1961). *The report of the scheduled areas and scheduled tribes commission, vol. 1, 1960–1961*. Delhi: Scheduled Areas and Scheduled Tribes Commission.

Dobhal, H. (2011). *Writings on human rights, law, and society in India: A combat law anthology: Selections from combat law, 2002–2010*. New Delhi: Socio Legal Information Cent.

Dove, M. R., Jonsson, H., & Aung-Thwin, M. (2011). Debate the art of not being governed: An anarchist history of upland southeast Asia by James C. Scott. *Bijdragen Tot de Taal- Land- En- Volkenkunde*, *167*(1), 86–99.

Drache, D. (1982). Harold Innis and Canadian capitalist development. *Canadian Journal of Political and Social Theory / Revue Canadienne de Théorie Politique et Sociale*, 6(1–2), 35–60.
Drayton, R. (1982). *Nature's government: Science, imperial Britain, and the improvement of the World*. New Haven: Yale University Press.
Dungdung, G. (2015). *Mission Saranda: A war for natural resource in India*. Ranchi: Deshaj Prakashan.
Dwivedi, R. (2006). *Conflict and collective action: The sardar sarovar project in India*. London and New York: Routledge.
Elwin, V. (1943). *The Baiga*. London: Oxford University Press.
Elwin, V. (1965). *Democracy in NEFA*. Shilong: North-East Frontier Agency.
Elwin, V., & Hivale, S. (1944). *Songs of the Maikal hills*. Bombay: Oxford University Press.
Elyachar, J. (2005). *Markets of dispossession: NGOs, economic development, and the state in Cairo*. Durham: Duke University Press.
Emmanuel, A. (1972). *Unequal Exchange: A study of the imperialism of trade*. New York: Monthly Review Press.
Engels, B., & Dietz, K. (2017). *Contested extractivism, society and the state: Struggles over mining and land*. London: Palgrave Macmillan Springer.
Enters, T., Durst, P. B., & Victor, M. (2000). *Decentralization and devolution of forest management in Asia and the Pacific*. RECOFTC Report No. 18 and RAP Publication 2000/1. Bangkok, Thailand.
Escobar, A. (1992). Imagining a post-development era? Critical thought, development and social movements. *Social Text, 31/32*, 20–56.
Escobar, A. (1996). Construction nature: Elements for a post-structuralist political ecology. *Futures*, *28*(4), 325–343.
Escobar, A. (2006a). An Ecology of difference: Equality and conflict in a glocalized world. *Focaal -European Journal of Anthropology*, *47*, 120–137. https://doi.org/10.3167/092012906780646415
Escobar, A. (2006b). Difference and conflict in the struggle over natural resources: A political ecology framework. *Development*, *49*(3), 6–13. Retrieved from http://www.palgrave-journals.com/development/journal/v49/n3/pdf/1100267a.pdf
Escobar, A. (2011). *Encountering development: The making and unmaking of the third world*. Princeton: Princeton University Press.
Escobar, A. (2014). Imagining a post-development era? Critical thought, development and social movements. *Social Text*, (31/32), 20–56.
Esteva, G. (1992). Development. In Wolfgang Sachs (Eds.), *The development dictionary: A guide to knowledge as power* (pp. 6–25). London: Zed Books Ltd.
Evans, P. (1995). *Embedded autonomy: States and industrial transformation*. Princeton, NJ: Princeton University Press.
Faubion, J. D. (Ed.). (2000). *Michel Foucault/power*. New York: The New Press.
Ferguson, J. (1990.). *The anti-politics machine: "Development", depoliticization, and bureaucratic power in lesotho*. Cambridge: Cambridge University Press.
Ferguson, J. (1994). The anti-politics machine: Development and bureaucratic power in Lesotho. *The Ecologists*, *24*(5), 176–181.
Filho, A. S., & Johnston, D. (2005). *Neoliberalism: A critical reader*. London: Pluto press.
Foster-Carter, A. (1976). From Rostow to Gunder Frank: Conflicting paradigms in the analysis of underdevelopment. *World Development*, *4*(3), 167–180.

Foucault, M. (1991). Governmentality. In M. Foucault, G. Burchell, C. Gordon, & P. Miller (Eds.), *The Foucault effect: Studies in Governmentality* (pp. 87–104). Chicago: University of Chicago Press.

Fox, R. G. (1971). *Kin, Clan, Raja, and Rule: State hinterland relations in preindustrial India*. Berkeley: University of California Press.

Frank, A. G. (1977). On so-called primitive accumulation. *Dialectical Anthropology*, 2(1), 87–106. https://doi.org/10.1007/BF00249476

Frank, A. G. (1978). Development of underdevelopment or underdevelopment of development in China. *Modern China*, *4*(3), 341–350.

Fraser, N. (1998). *Social justice in the age of identity politics: Redistribution, recognition, participation* (No. FS I, pp. 98–108). Berlin. Retrieved from https://www.ssoar.info/ssoar/bitstream/handle/document/12624/ssoar-1998-fraser-social_justice_in_the_age.pdf?sequence=1

Freire, P. (2000). *Pedagogy of the oppressed*. New York: Continuum International Publishing Group Inc.

Friedman, J. (1974). Marxism, structuralism, and vulgar materialism. *Man, New Series*, *9*(3), 444–469.

Gadgil, M., & Guha, R. (1992). *The fissured land: An ecological history of India*. Delhi: Oxford University Press.

Gadgil, M., & Guha, R. (2013). *Ecology and equity: The use and abuse of nature inc Contemporary India*. London: Routledge.

Geertz, C. (1977). *The interpretation of cultures*. New York: Perseus Books Group.

Gell, A. (1997). Exalting the king and obstructing the state: A political interpretation of royal ritual in Bastar District, Central India. *Journal of the Royal Anthropological Institute*, (3), 433–450.

George, A. (2009). The paradox of mining and development. In N. Sundar (Ed.), *Legal grounds: Natural resources, identity, and the law in Jharkhand* (pp. 157–188). New Delhi: Oxford University Press.

George, A. (2014). *Status of adivasis/indigenous peoples mining series*. New Delhi: Aakar Books.

Ghani, A., Lockhart, C., & Carnahan, M. (2005). *Closing the sovereignty gap: An approach to state building* (No. 253). London.

Giddens, A. (1979). *Central problems in social theory: Action, structure and contradiction in social analysis*. Berkeley: University of California Press.

Gidwani, V. K. (2008). *Capital, interrupted: Agrarian development and the politics of work in India*. Minnesota: University of Minnesota Press.

Ginger, C., Emery, M. R., Baumflek, M. J., & Putnam, D. E. (2012). Access to natural resources on private property: Factors beyond right of entry. *Society & Natural Resources*, *25*(7), 700–715. https://doi.org/10.1080/08941920.2011.633596

Glassman, J. (2006). Primitive accumulation, accumulation by dispossession, accumulation by 'extra-economic'means. *Progress in Human Geography*, *30*(5), 608–625. https://doi.org/10.1177/0309132506070172

Gleditsch, N. P., & Urdal, H. (2002). Ecoviolence? Links between population growth, environmental scarcity and violent conflict in Thomas Homer-Dixon's work. *Journal of International Affairs*, *56*(1), 283–302.

GoI (Government of India). (1996). *Panchayat (Extension to Scheduled Areas) Act, 1996*. New Delhi. https://www.mha.gov.in/sites/default/files/PESAAct1996_0.pdf

Gooptu, N., & Parry, J. (Eds.). (2017). *Persistence of poverty in India*. London: Routledge.

Gordon, C. (1991). *Governmental Rationality: An Introduction. The Foucault Effect: Studies in Governmentality*. https://doi.org/10.1146/annurev-micro-090110-102957

Government of India. (1982). *Report of the committee on forest and tribals in India*. Delhi.

Grove, R. (1990). Colonial conservation, ecological hegemony and popular resistance: Towards a global synthesis. In J. MacKenzie (Ed.), *In imperialism and the natural world* (pp. 15–50). Manchester: Manchester University Press.

Gudynas, E. (2010). *The new extractivism of the 21st century: Ten urgent theses about extractivism in relation to current South American progressivism*. Washington, DC: Americas Program Report.

Guha, R. (1983). *Elementary aspects of peasant insurgency in colonial India*. Delhi: Oxford University Press.

Guha, R. (1989). *The unquiet woods: Ecological change and peasant resistance in the Himalaya*. New Delhi: Oxford University Press.

Guha, R. (1997). *Dominance without hegemony: History and power in colonial India*. Cambridge: Harvard University Press.

Guha, R. (2010). Adivasis: Unacknowledged victims. *Outlook*. Retrieved from https://www.outlookindia.com/website/story/unacknowledged-victims/265069

Guizzo, D., & Vigo de Lima, I. (2015). Polanyi and Foucault on the issue of market in classical political economy: Complementary approaches to the radical theory of social control. *Review of Radical Political Economics*, 1–14. https://doi.org/10.1177/0486613415621745

Gupta, A. (2012). *Red Tape: Bureaucracy, structural violence, and poverty in India*. London: Duke University Press.

Gupta, A., & Ferguson, J. (2002). Spatializing states: Toward an ethnography of neoliberal governmentality. *American Ethnologist*, *29*(4), 981–1002.

Habib, I (1995). *Essays in Indian history: Towards a marxist perception*. New Delhi: Tulika

Haesbaert, R. (2013). A global sense of place and multi-territoriality notes for dialogue from a peripheral point of View. In *Spatial politics essays for Doreen Massey* (pp. 146–157). London: John Wiley and Sons Ltd.

Hall, S., Morley, D., & Chen, K.-H. (1996). Stuart Hall: Critical dialogues in cultural studies. *Comedia*. https://doi.org/10.4324/9780203993262

Hammond, L. (2011). Governmentality in motion: 25 years of Ethiopia's experience of famine and migration policy. *Mobilities*, *6*(3), 415–432.

Hansen, T. H., & Stepputat, F. (2001). *States of imagination: Ethnographic explorations of the postcolonial state*. London: Duke University Press.

Haraway, D. (2013). *Simians, cyborgs, and women: The reinvention of nature*. London: Routledge.

Hardiman, D. (1987). *The coming of the devi: Adivasi assertion in western India*. London: Oxford University Press.

Hardin, G. (1968). The tragedy of the commons. *Science*, *162*(3859), 1243–1248.

Harris, L. M. (2017). Political ecologies of the State: Recent interventions and questions going forward. *Political Geography*, *58*, 90–92.

Harriss, J., & Jeffrey, C. (2013). Depoliticizing injustice. *Economy and Society*, *42*(3), 507–520.

Harriss-White, B. (1997). *Informal economic order, shadow states, private status states, states of last resort and spinning states: A speculative discussion based on*

south Asian (No. S06). *Queen Elizabeth House, Working Paper Series, ...* Retrieved from http://www.cs.odu.edu/~dlibug/ups/rdf/remo/qeh/qehwps/qehwsp06.pdf
Harriss-White, B. (2003). *India working: Essays on society and economy*. London: Cambridge University Press.
Harriss-White, B. (2014). *Indian capitalism in development*. London: Routledge.
Harriss-White, B., & Prakash, A. (2013). *Social discrimination in India: A case for economic citizenship* (Oxfam India Working Paper Series No. 5). https://doi.org/10.1017/CBO9781107415324.004
Harvey, D. (2004). The "New" imperialism: Accumulation by dispossession. *Socialist Register*, *40*(1), 63–87. Retrieved from http://socialistregister.com/index.php/srv/article/view/5811
Harvey, D. (2005). *A brief history of Neoliberalism*. Oxford: Oxford University Press.
Harvey, D. (2007). Neoliberalism as creative destruction. *The ANNALS of the American Academy of Political and Social Science*, *610*(1), 21–44. https://doi.org/10.1177/0002716206296780
Harvey, D. (2014). *Seventeen Contradictions and the End of Capitalism*. New York, NY: Oxford University Press.
Hebber, R. (2011). *Ecology, equality and freedom: The engagement with self-rule in Jharkhand*. Navi Mumbai: Earthworm Books Pvt. Ltd.
Hefner, R. W. (1990). *The political economy of mountain Java: An interpretive history*. Berkeley: University of California Press.
Heller, P. (1996). Social capital as a product of class mobilization and state intervention: Industrial workers in Kerala. India. *World Development*, *24*(6), 1055–1071.
Heredia, R. C. (2002). Interrogating integration: The counter cultural tribal other. *Economic & Political Weekly*, *37*(52), 5174–5178.
Hill, J. K. W. (2014). Agriculture, irrigation and ecology in adivasi villages in Jharkhand: Why control and ownership over natural resources matter. *Journal of Adivasi and Indigenous Studies (JAIS)*, *1*(1), 43–61.
Hoare, Q., & Nowell Smith, G. (Eds.). (1971). *Selection from the prison notebooks of Antonio Gramsci*. London: Lawrence & Wishart.
Hoben, A. (1995). Paradigms and politic: The cultural construction of environmental policy in Ethiopia. *World Development*, *23*(6), 1007–1021.
Hobsbawm, E. (1973). Peasants and politics. *Journal of Peasant Studies*, *1*(1), 3–22.
Humphreys, M. (2005). Natural resources, conflict, and conflict resolution uncovering the mechanisms. *Journal of Conflict Resolution*, *49*(4), 508–537.
Ioris, A. A. R. (2015). *The political ecology of the state: The basis and the evolution of the environmental statehood*. London and New York: Routledge.
Jayal, N. (2000). *Democracy and the state: Welfare, secularism and development in contemporary India*. Delhi: Oxford University Press.
Jessop, B. (1999). Narrating the future of the national economy and the national state? Remarks on remapping regulation and reinventing governance. In G. Steinmetz (Ed.), *In state/culture: State formation after the cultural turn* (pp. 378–405). Ithaca, NY: Cornell University Press.
Jessop, B. (2009). Redesigning the state, reorienting state power, and rethinking the state. In J. Jenkins & K. Leicht (Eds.), *Handbook of politics: State and society in global perspective* (pp. 41–61). New York: Springer.
Jessop, B. (2016). *The state: Past, present, future*. Cambridge: Polity Press.

Jewitt, S. (2008). Political ecology of Jharkhand conflicts. *Asia Pacific Viewpoint, 49*(1), 68–82. https://doi.org/10.1111/j.1467-8373.2008.00361.x

Jewitt, S., & Kumar, S. (2000). A political ecology of forest management: Gender and silvicultural knowledge in Jharkhand, India. In P. A. Stott & S. Sullivan (Eds.), *Political ecology: Science, myth and power*. New York: Arnold.

Jha, J. C. (1964). *The Kol insurrection of Chota-Nagpur*. Calcutta: Thacker, Spink & Co.

Jodha, N. S. (1990). Rural common property resources: Contributions and crisis. *Economic and Political Weekly, 25*(26), A65–A78.

Kaltwasser, R. C. (2011). Toward post neoliberlarism in Latin America. *Latin American Research Review, 46*(2), 225–234.

Kapila, K. (2008). The measure of a tribe: The cultural politics of constitutional reclassification in north India. *Journal of the Royal Anthropological Institute, 14*, 117–134.

Kapoor, D. (2012). Human rights as paradox and equivocation in contexts of adivasi (original dweller) dispossession in India. *Journal of Asian and African Studies, 47*(4), 404–420. https://doi.org/10.1177/0021909612438092

Karan, P. P. (1957). Iron mining industry in Singhbhum-Mayurbhanj region of India. *Economic Geography, 33*(4), 349–361.

Karlsson, B. G. (2004). Beyond integration: Indigenous assertion in India. Retrieved from https://iias.asia/sites/default/files/IIAS_NL35_08.pdf

Karmakar, K. G. (2002). *The silenced drums: A review of tribal economic development*. Northern Book Centre. Retrieved from https://books.google.com/books?id=QI93ntXUA4sC&pgis=1

Kaviraj, S. (1988). A critique of the passive revolution. *Economic and Political Weekely, 23*(45/47), 2429–2444.

Kishwar, M. (1987). Toiling without rights: Ho women of Singhbhum. *Economic and Political Weekly, 22*(3), 95–101.

Kohli, A. (1987). *The state and poverty in India: The politics of reform*. New York: Cambridge University Press.

Kohli, A. (2004). *State-directed development: Political power and industrialization in the global Periphery*. Cambridge: Cambridge University Press.

Kohli, A. (Ed.). (2014). *India's democracy: An analysis of changing state-society relations*. New Jersy: Princeton University Press.

Kothari, R. (1990). *State against democracy: In search of human governance*. New Delhi: Aspect.

Krancberg, S. (1986). Common sense and philosophy in Gramsci's prison notebooks. *Studies in Soviet Though, 32*(2), 163–181.

Kumar, D., & Mishra, N. R. (2017). Observing developmental intervention in mineral-rich adivasi subsistence economy: An ethnographic account from West Singhbhum (Jharkhand). *Contemporary Voice of Dalit, 9*(1), 73–86.

Kumar, D., & Puthumattathil, A. (2018). A critique of development in India's predominantly adivasi regions with special reference to the Hos of India's Jharkhand. *Contemporary Voice of Dalit, 10*(1), 10–27.

Kumar, S. (1983). *Report of the commissioner for Scheduled castes and Scheduled Tribe (Twenty Seventh Report) Part 1, 1978–81*. Delhi.

Kumar, S., & Corbridge, S. (2002). Programmed to fail? Development projects and the politics of participation. *The Journal of Development Studies, 39*(2), 73–103. https://doi.org/10.1080/00220380412331322761

Kuper, A. (2005). *The reinvention of primitive society: Transformations of a Myth*. London: Routledge.

Lane, R. E., & O'Connor, M. (1995). Is capitalism sustainable? Political economy and the politics of ecology. *Political Psychology*, *16*(4), 869. https://doi.org/10.2307/3791899

Laski, H. J. (2010). *A grammar of politics: Works of Harold J. Laski*. London: Taylor & Francis Group.

Latour, B. (1998). To modernize or to ecologize? That's the question. In N. Castree & B. W. Braun (Eds.), *Remaking reality: Nature at the millenium* (pp. 221–242). London: Routledge.

Le Billon, P. (2011). The political ecology of war: Natural resources and armed conflicts. *Political Geography*, *20*(5), 561–584.

Leach, M., & Mearns, R. (1996). Environmental change and policy: Challenging received wisdom in Africa. In M. Leach & R. Mearns (Eds.), *The lie of the land: Challenging received wisdom on the African Environment* (pp. 1–33). London: The International African Institute.

Lebowitz, M. (2002). Karl Marx: The need of capital vs the needs of human beings. In D. Dowd (Ed.), *Understanding capitalism: Critical analysis from Karl Marx to Amartya Sen* (pp. 17–36). London: Pluto press.

Lebowitz, M. (2009). The path to human development: Capitalism or socialism. *Monthly Review*. Retrieved from https://monthlyreview.org/2009/02/01/the-path-to-human-development-capitalism-or-socialism/

Lele, S. (2004). *Decentralising governance of natural resources in India: A review, a report submitted to UNDP Dryland Development Centre, Nairobi*. Bangalore: Centre for Interdisciplinary Studies in Environment Development.

Lemke, T. (2010). Foucault, governmentality, and critique. *Rethinking Marxism A Journal of Economics, Culture & Society*, *14*(3), 49–64. Retrieved from https://www.tandfonline.com/doi/abs/10.1080/089356902101242288?journalCode=rrmx20

Lemm, V., & Vatter, M. (Eds.). (2014). *The government of life: Foucault, biopolitics, and neoliberalism*. New York: Fordham University Press.

Lenin, V. I. (1999). *Imperialism, the Highest Stage of Capitalism*. Resistance Books.

Lerner, D. (1958). *The passing of traditional society: Modernizing the middle east*. New York: Free Press of Glencoe.

Levien, M. (2011). Special economic zones and accumulation by dispossession in India. *Journal of Agrarian Change*, *11*(4), 454–483. https://doi.org/10.1111/j.1471-0366.2011.00329.x

Levien, M. (2015). From primitive accumulation to regimes of dispossession: Six theses on India's land question. *Economic & Political Weekly*, *1*(22), 146–157.

Levien, M. (2018). *Dispossession without development: Land grabs in neoliberal India*. London: Oxford University Press.

Li, T. M. (1999). Compromising power: Development, culture, and rule in Indonesia. *Cultural Anthropology*, *14*(3), 295–322. https://doi.org/10.1525/can.1999.14.3.295

Li, T. M. (2000). *Articulating indigenous identity in Indonesia: Resource politics and tribal slot* (No. WP 00-7). *Berkeley Workshop on Environmental Politics*.

Li, T. M. (2005a). Beyond "the state" and failed schemes. *American Anthropologist*, *107*(3), 383–394. https://doi.org/10.1525/aa.2005.107.3.383

Li, T. M. (2005b). *Transforming the Indonesian uplands*. London: Routledge. Retrieved from https://books.google.com/books?id=6jMBnvUC--YC&pgis=1
Li, T. M. (2007). *The will to improve: Governmentality, development, and the practice of politics*. Durham: Duke University Press.
Li, T. M. (2007b). Ideas / idees governmentality. *Anthropologica*, *49*(2), 275–281.
Li, T. M. (2010). Indigeneity, capitalism, and the management of dispossession. *Current Anthropology*, *51*(3), 385–414. https://doi.org/10.1086/651942
Li, T. M. (2012). To make live or let die? Rural dispossession and the protection of surplus populations. *Antipode*, *41*, 66–93. https://doi.org/10.1002/9781444397352.ch4
Li, T. M. (2014). *Land's end: Capitalist relations on an indigenous frontier*. Durham: Duke University Press. https://doi.org/10.1215/9780822376460
Limb, M., & Dwyer, C. (2001). *Qualitative methodologies for geographers: Issues and debates*. London: Arnold.
Ludden, D. (2003). Investing in nature around Sylhet: An excursion into geographical history. *Economic and Political Weekly*, *38*(48), 5080–5088. Retrieved from http://www.jstor.org/stable/4414346
Lund, C. (2008). *Local politics and the dynamics of property in Africa*. New York: Cambridge University Press.
Mann, G. (2009). Should political ecology be Marxist? A case for Gramsci's historical materialism. *Geoforum*, *40*(3), 335–344. https://doi.org/10.1016/j.geoforum.2008.12.004
Mann, M. (1984). The autonomous power of the state: Its origins, mechanisms and results. *European Journal of Sociology*, *25*(2), 185–213.
Mann, R. S. (Ed.). (1981). *Nature man spirit: Complex in tribal India*. Delhi: Concept Publishing Company.
Marshall, T. H. (1950). *Citizenship and social class*. Cambridge: Cambridge University Press.
Marshall, T. H. (1987). *Citizenship and social class: And other essays*. London: Pluto press.
Martinez-Alier, J. (1994). Ecological distribution conflicts and indicators of sustainability 1. *Revista Ibero-Americana de Economia Ecologica*, *1*(1), 21–31.
Martinez-Alier, J. (2014). The environmentalism of the poor. *Geoforum*. Elsevier Ltd.
Marx, K. (1859). *A contribution to the critique of political economy*. Moscow: Progress Publishers.
Marx, K. (1977). *Capital: A critique of political economy* (Vol. I). Trans. B. Fowkes. New York: Penguin.
Marx, K. (1986). *Capital: A critique of political economy* (Vol 1). Moscow: Progress Publishers.
Maxwell, J. A. (2012). *Qualitative research design: An interactive approach*. London: SAGE Publications.
Mazaheri, N. (2016). *Oil booms and business Busts: Why resource wealth hurts rntrepreneurs in the developing world*. Oxford: Oxford University Press.
Mbembe, A. (1992). The binality of power and the asthetics of vulgarity in the post colony. *Public Culture*, *4*(2), 1–30.
Mildavin, J. S. S. (2007). The politics of transition: Critical political ecology, classical economics, and ecological modernization theory in China. In J. Robinson & K. R. Cox (Eds.), *The SAGE handbook of political geography* (pp. 247–262). London: SAGE Publications.

Mishra, A., & Paty, C. K. (2010). *Tribal movements in Jharkhand 1857–2007*. New Delhi: Concept Publishing Company.

Mitchell, T. (1991). The limits of the state: Beyond statist approaches and their critics. *American Political Science Review*, *85*(1), 77–96.

Mitchell, W. J. T. (Ed.). (1994). *Landscape and power* (2nd ed.). Chicago: University of Chicago Press.

Mitra, S. K. (2002). *Power, protest and participation: Local Elites and development in India*. London: Routledge.

Moore, D. S. (1993). Contesting terrain in Zimbabwe eastern highlands: Political ecology, ethnography, and peasant resource struggles. *Economic Geography*, *69*(4), 380–401. https://doi.org/10.2307/143596

Moore, D. S. (2005). *Suffering for territory: Race, place, and power in Zimbabwe*. Durham: Duke University Press.

Moose, D. (2007). *Power and the durability of poverty: A critical exploration of the links between culture, marginality and chronic poverty* (No. 107).

Mukesh, A. S. R. P. (2012a, April). Saranda bleeds in illegal ore hunt. *The Telegraph*. Retrieved from http://www.telegraphindia.com/1120425/images/25zzjhanumberbig.jpg

Mukesh, A. S. R. P. (2012b, May). Showcause glare on Saranda pilferage. *The Telegraph*. Retrieved from http://www.telegraphindia.com/1120530/jsp/frontpage/story_15548113.jsp#.WB2awtV97IV

Mullick, S. B. (2004). Conflict and resolution between indigenous Peoples and the state of India. In A. L. Tamayo, E. R. Alcantara, M. Malanes, R. de Chavez, & R. Sidchogan-Batani (Eds.), *Reclaiming balance: Indigenous peoples, conflict resolution and sustainable development* (pp. 153–172). Philippines: Tebtebba Foundation.

Muradian, R., Walter, M., & Alier, M. (2016). Hegemonic transitions and global shifts in social metabolosm: Implications for resource- rich contries. Introduction to the special section. *Global Environment Change*, *22*(3), 559–567.

NABARD. (2014). *District agriculture plan: West Singhbhum district, 2008-09 to 2011-12*. Ranchi: NABARD Consultancy Services, Jharkhand Regional Office. Retrived from http://nrcddp.org/resources_data/Dis100142.pdf

Naik, S. C., & Tiwari, T. N. (2007). *Society and environment*. New Delhi: Oxford & IBH, Ed.

Nandy, A. (2002). *The romance of the state: And the fate of dissent in the tropic*. New Delhi: Oxford University Press.

Nandy, A., & Visvanathan, S. (1990). Modern medicine and its non modern critics. In F. A. Marglin & S. A. Margin (Eds.), *Dominating knowledge: Development, culture, and resistance* (pp. 145–184). London: Oxford University Press.

Nathan, D., & Kelker, G. (2003). Civilisational change: Markets and privatisation among Indigenous Peoples. *Economic and Political Ecology*, *38*(20), 1955–1968.

Nathan, D., & Xaxa, V. (2012). *Social exclusion and adverse inclusion: Development and deprivation of adivasis in India*. New Delhi: Oxford University Press.

Nehru, J. L. (1946). *The discovery of India*. New Delhi: Penguin UK.

Neumann, R. P. (2008). Probing the (in)compatibilities of social theory and policy relevance in Piers Blaikie's political ecology. *Geoforum*, *39*(2), 728–735.

Nightingale, A. (2002). Can social theory adequately address nature- society issues? Do political ecology and science studies in geography incorporate ecological change? In *Third International Critical Geography Conference* (pp. 1–20). Bekescsaba Hungary.

Nilsen, A. G. (2010). *Dispossession and resistance in India: The river and the rage*. London: Routledge.

Nilsen, A. G. (2013). Against the current, from below: Resisting dispossession in the Narmada valley, India. *Journal of Poverty*, *17*(4), 460–492.

Nilsen, A. G. (2016). Democratic struggles in the Bhil heartland: Historical trajectories and contemporary scenarios. In U. Chandra & D. Taghioff (Eds.), *Staking claims: The politics of social movements in contemporary rural India* (pp. 31–63). New Delhi: Oxford University Press.

Nilsen, A. G., & Roy, S. (2013). Introduction: Reconceptualizing Subaltern Politics in Contemporary India. In A. G. Nilsen & S. Roy (Eds.), *New Subaltern Politics: Reconceptualizing Hegemony and Resistance in Contemporary India* (pp. 1–28). Delhi: Oxford Academic.

Norgaard, R. B. (2006). *Development betrayed: The end of progress and a co-evolutionary revisioning of the future*. London: Routledge.

Nunan, F. (2015). *Understanding poverty and the environment: Analytical frameworks and approaches*. London: Routledge.

Oliver-Smith, A. (2006). Displacement, resistance and the critique of development: From the grassroots to the global. In C. De Wet (Ed.), *Development-induced displacement: Problems, policies, and people* (pp. 141–179). London and New York: Berghahn Books.

Omeje, K. (Ed.). (2008). *Extractive economies and conflicts in the global south: Multi-regional perspectives on rentier politics*. London: Ashgate.

Omeje, K. (2013). *Extractive economies and conflicts in the global south: Multi-regional perspectives on rentier politics*. Ashgate Publishing, Ltd. Retrieved from https://books.google.com/books?id=_lKUXro-4wsC&pgis=1

Padel, F. (2014). *Interview: Anthropologist Felix Padel on life and struggle in Niyamgiri*. Retrieved from http://www.indiaresists.com/interview-anthropologist-felix-padel-on-life-and-struggle-in-niyamgiri/

Padel, F. (2016). In the name of sustainable development: Genocide masked as tribal development. In *First citizen: Studies on adivasis, tribals and indigenous peoples in India* (pp. 159–178). New Delhi: Oxford University Press.

Padel, F., Dandekar, A., & Unni, J. (2013). *Ecology, economy: Quest for a socially informed connection*. Hyderabad: Orient Blackswan.

Padel, F., & Das, S. (2010). Cultural genocide and the rhetoric of sustainable mining in East India. *Contemporary South Asia*, *18*(3), 333–341. https://doi.org/10.1080/09584935.2010.503871

Panda, P. (2005). *Traditional political system of the Binjhals*. New Delhi: Northern Book Centre.

Parajuli, P. (1996). Ecological ethnicity in the making: Developmentalist hegemonies and emergent identities in India. *Identities*, *3*(1–2), 14–59. https://doi.org/10.1080/1070289X.1996.9962551

Parajuli, P. (2010). At home in Prescott: Confluence of ten streams in my journey as sustainability educator. *Journal of Sustainability Education*, Inaugural Issue, May 2010. http://www.journalofsustainabilityeducation.org/wordpress/content/at-home-in-prescott-confluence-of-streams-in-my-journey-as-an-interdisciplinary-sustainability-educator_2010_05/

Parajuli, P., & Kothari, S. (1999). Struggling with autonomy: Lessons from local governance.

Parsons, T. (1991). *The social system*. London: Routledge.

Passi, A. F. (2003). Territory. In J. A. Agnew, K. Mitchell, & G. Toal (Eds.), *A companion to political geography* (pp. 109–123). USA: Blackwell Publishing.

Pati, B. (2011). *Adivasis in colonial India: Survival, resistance and negotiation*. New Delhi: Orient Blackswan.

Paulson, S., & Gezon, L. L. (Eds.). (2005). *Political ecology across space, scales, and social groups*. London: Rutgers University Press.

Peet, R., & Hartwick, E. (2009). *Theories of development: Contention, arguements, alternatives*. New York: The Guilford Press.

Peet, R., & Watts, M. (2004). *Liberation ecologies: Environment, development and social movements*. New York: Routledge. Retrieved from https://books.google.com/books?id=lgSdlm7XExQC&pgis=1

Peet, R., & Watts, M. (2016). Introduction: Development theory and environment in an age of market triumphalism. *Economic Geography*, *69*(3), 227–253.

Peluso, N. Lee, & Vandergeest, P. (2001). Genealogies of the political rights forest and customary in Indonesia, Malaysia, and Thailand. *The Jounral of Asian Studies*, *60*(3), 761–812.

Peluso, N. L. (2009). Rubber erasures, rubber producing rights: Making racialized territories in west kalimantan, Indonesia. *Development and Change*, *40*(1), 47–80.

Peluso, N. L., & Lund, C. (2011). New frontiers of land control: Introduction. *Journal of Peasant Studies*, *38*(4), 667–681. https://doi.org/10.1080/03066150.2011.607692

Peluso, N. L., & Vandergeest, P. (2011). Political ecologies of war and forests: Counter insurgencies and the making of national natures. *Annals of the Association of American Geographers*, *101*(3), 587–608. https://doi.org/10.1080/00045608.2011.560064

Peluso, N. L., & Watts, M. (2001). Violent environments. In N. L. Peluso & M. Watts (Eds.), *Violent environments* (p. 453). New York: Cornell University Press.

Pieterse, J. N. (2000). After post development. *Third World Quarterly*, *21*(2), 175–191.

Polanyi, K. (1944). *The great transformation: The political and economic origins of our time*. Boston: Beacon Press.

Poma, A., & Gravante, T. (2015). Analyzing resistance from below: A proposal of analysis based on three struggles against dams in Spain and Mexico. *Capitalism Nature Socialism*, *26*(1), 59–76.

Prabhu, P. (2004). PESA and the illusions of tribal self-governance. In Harsh Dobhal (Eds.), *Writings on human rights, law and society in India: a combat law anthology* (pp. 570–577). Delhi: Human Rights Law Network.

Prakash, A. (1999). Contested discourses: Politics of ethnic identity and autonomy in the Jharkhand region of India. *Altenatives: Global, Local, Political*, *24*(4), 461–496.

Prakash, A. (2011). Politics, development and identity Jharkhand, 1991–2009. In D. J. Rycroft & S. Dasgupta (Eds.), *The politics of belonging in India: Becoming adivasi* (1st ed., pp. 175–190). New York: Routledge.

Prasad, A. (2016). Adivasis' and the trajectories of political mobilisation in Contemporary India. In M. Radhakrishna (Ed.), *First citizen: Studies on adivasis, tribals and Indigenous Peoples in India* (pp. 307–336). New Delhi: Oxford University Press.

Prasad. (2008). *Environment, development and society in contemporary India: An introduction*. New Delhi: Macmillan India.

Puthumattathil, A. (2014). *Colonialism and racism uninterrupted: Evidence from India with special reference to the Hos of Jharkhand.* (Unpublished doctoral dissertation). Ghent University, Faculty of Political and Social Sciences, Ghent, Belgium.

Rangan, H. (1997). Indian environmentalism and the question of the state: Problems and prospects for sustainable developmen. *Environment and Planning A: Economy and Space*, *29*(12), 2129–2143.

Rasmussen, M. B., & Lund, C. (2018). Reconfiguring frontier spaces: The territorialization of resource control. *World Development*, *101*(C), 388–399.

Ratnagar, S. (2003). Our tribal past. *Social Scientists*, *31*(1–2), 17–36.

Reddy, G. R., & Haragopal, G. (1985). The pyraveekar: "The fixer" in rural India. *Asian Survey*, *25*(11), 1148–1162.

Ribot, J. C., & Peluso, N. Lee. (1999). A theory of access. *Rural Sociology*, *68*(2), 153–181. https://doi.org/10.1111/j.1549-0831.2003.tb00133.x

Roa, A. (2009). *Field notes on democracy: Listening to grasshoppers.* Chicago: Haymaker Books.

Robbins, P. (2012). *Political Ecology A critical Introduction* (2nd ed.). Sussex: Wiley-Blackwell.

Robbins, P. (2014). Cries along the chain of accumulation. *Geoforum*, *54*, 233–235. https://doi.org/10.1016/j.geoforum.2012.12.007

Robertson, M. (2015). Environmental governance: Political ecology and the state. In T. Perreault, G. Bridge, & J. McCarthy (Eds.), *The Routledge handbook of political ecology* (pp. 457–466). London: Routledge.

Rosin, R. T. (1993). The tradition of groundwater irrigation in northwestern India. *Human Ecology*, *21*(1), 51–86.

Ross, M. L. (2004). What do we know about natural resources and civil war? *Journal of Peace Research*, *41*(3), 337–356.

Rostow, W. W. (1991). *The stages of economic growth: A non-communist manifesto.* Cambridge: Cambridge University Press.

Roy, S. (2012). *Madhu Koda loot raj.* New Delhi: Prabhat Publishers.

Roy, S. C. (1970). *The mundas and their country.* Bombay: Asia Publishing House.

Rudolph, L. I., & Rudolph, S. H. (1987). *In pursuit of lakshmi: The political economy of the Indian state.* Chicago: University of Chicago Press.

Rycroft, D. J., & Dasgupta, S. (Eds.). (2011). *The politics of belonging in India: Becoming adivasi.* London: Routledge.

Sachchidananda. (1972). The tribal situation in Bihar. In K.S. Singh (Ed.), *Tribal situation in India* (pp. 169–183). Shimla: Indian Institute of Advanced Studies.

Sachs, W. (Ed.). (2010). *The development dictionary: A guide to knowledge as power.* New York: Zed Books Ltd.

Sack, R. D. (1983). Human territoriality: A theory. *Annals of the Association of American Geographers*, *73*(1), 55–74.

Sack, R. D. (1986). *Human territoriality: Its theory and history.* Cambridge: Cambridge University Press.

Sader, E. (2009). Postneoliberalism in latin America. *Development Dialogue*, *51*(1), 171–179.

Saha, S. K. (1986). Historical premises of India's tribal problem. *Journal of Contemporary Asia*, *16*(3), 274–319.

Sahoo, D., & Kumar, D. (2017a). From Polanyi to Gramsci and Foucault: Looking at polanyian concepts through the lens of neoliberal cultural politics of development in India. *Eastern Anthropologist*, *70*(3–4), 427–437.

Sahoo, D., & Kumar, D. (2017b). From Polanyi to Gramsci and Foucault: Looking at Polanyian concepts through the lenses of Neoliberal cultural politics of development in India. In *A Great Transformation? Global Perspectives on Contemporary Capitalisms* (pp. 1–14). Linz: Karl Polanyi Institute of Political Economy. Retrieved from https://www.jku.at/conferences/content/e290365/.../Sahoo_Kumar_GreatTrans.pdf

Sahu, M. (1985). *The Kolhan under British rule*. Jamsedpur: Utkal Book Agency.

Sangampam, S. N. (1986). The state-society relationship in peripheral countries: Critical notes on the dominant paradigms. *The Review of Politics*, *48*(4), 596–619.

Santos, B. D. S. (2012). Public sphere and epistemologies of the south. *Africa Development*, *37*(1), 43–67.

Santos, B. D. S. (2015). *Epistemologies of the south: Justice against epistemicide* (Vol. 17). Routledge. Retrieved from https://books.google.com/books?id=P2nvCgAAQBAJ&pgis=1

Sanyal, K. (2007). *Rethinking capitalist development: Primitive accumulation, governmentality and post-colonial capitalism*. London: Routledge.

Saraswati, B. (1991). *Tribal thought and culture: Essays in honour of Surajit Chandra Sinha* (B. Saraswati, Ed.). Delhi: Concept Publishing Company.

Sareen, S. (2015). *Governing the conflicted commons: Authorising resource access in the Indian tribal Belt*. (Unpublished doctoral dissertation). Department of Food and Resource Economics, PhD School of the Faculty of Science, University of Copenhagen, Copenhagen.

Sareen, S., & Nathan, I. (2018). Under what conditions can local government nurture indigenous people's democratic practice? A case study of two Ho village assemblies in Jharkhand. *The Journal of Development Studies*, *54*(8), 1354–1373.

Sartre, J. P. (2005). *Colonialism and neocolonialism*. London: Routledge.

Sauer, S. (2012). Land and territory: Meaning of land between modernity and tradition. *Agararian South: Journal of Political Eonomy*, *1*(1), 85–107.

Savyasaachi. (2016). Primitive accumulation, labour, and the making of "schduled tribe", "indigeneous", and adivasi sensibity. In M. Radhakrishna (Ed.), *First citizen: Studies on adivasis, tribals and indigenous peoples in India* (pp. 53–76). New Delhi: Oxford University Press.

Schlosberg, D. (2007). *Defining environmental justice: Theories, movements and nature*. New York: Oxford University Press.

Schneider, A. E. (2011). What shall we do without our land? Land Grabs and Resistance in Rural Cambodia. *Paper presented at the International Conference on Global Land Grabbing, April 6–8, 2011, Land Deals Politics Initiative (LDPI)*, University of Sussex.

Schneider, A. E. (2011). What shall we do without our land? Land grabs and resistance in rural Cambodia. In *Global Land Grabbing* (pp. 1–36). Sussex.

Scott, J. C. (1989). Everyday forms of resistance. *The Copenhagen Journal of Asian Studies*, *89*(4), 33–69. https://doi.org/10.1111/j.1469-0691.2011.03558.x/pdf

Scott, J. C. (1998a). *Seeing like a state: How certain schemes to improve the human condition have failed*. New Haven: Yale University Press.

Scott, J. C. (2010). *The art of not being governed: An anarchist history of upland south east Asia*. New Delhi: Orient Blackswan.

Scott, J. C. (1976). *The moral economy of the peasant: Rebellion and subsistence in southeast Asia*. New Haven and London: Yale University Press.

Scott, J. C. (1985). *Weapons of the weak: Everyday forms of peasant resistance*. New Haven: Yale University Press.
Scott, J. C. (1998b). Freedom and freehold: Space, people and state simplification in southeast Asia. In D. Kelly & A. Reid (Eds.), *The idea of freedom in east and southeast Asia* (pp. 37–64). Cambridge: Cambridge University Press.
Sen, A. (1999). *Development as freedom*. London: Oxford University Press.
Sen, A. K. (1999). *Wilkinson's rules: Context, content and ramifications*. Chaibasa: Tata College.
Sen, A. K. (2011). *Representing tribe: The Ho of Singhbhum under colonial rule*. New Delhi: Concept Publishing Company.
Sen, A. K. (2018). *Indigeneity, landscape and history: Adivasi self fashioning in India*. London: Routledge.
Shade, L. (2015, August). Sustainable development or sacrifice zone? Politics below the surface in post-neoliberal Ecuador. *The Extractive Industries and Society*. http://dx.doi.org/10.1016/j.exis.2015.07.004
Shah, A. (2007). 'Keeping the state away': Democracy, politics, and the state in India's Jharkhand. *Journal of Royal Anthropological Institute*, *13*(1), 129–145.
Shah, A. (2010). *In the shadows of the state: Indigenous politics, environmentalism, and insurgency in Jharkhand, India*. Durham: Duke University Press.
Shah, A. (2013). The agrarian question in a maoist guerrilla zone: Land, labour and capital in the forests and hills of Jharkhand. *Journal of Agrarian Change*, *13*(3), 424–450.
Shah, A. (2017). Ethnography? Participant observation, a potentially revolutionary praxis. *Hau: Journal of Ethnographic Theory*, *7*(1), 45–59.
Shah, M. B. (2013). *First report on illegal Mining of Iron and Manganese Ores in the State of Jharkhand*. Retrieved from http://mines.nic.in/writereaddata/Upload File/First Report on Jharkhand Vol - I.pdf
Sharan, R. (2009). Alienation and restoration of tribal land in Jharkhand. In N. Sundar (Ed.), *Legal grounds: Natural resources, identity, and the law in Jharkhand* (pp. 82–112). New Delhi: Oxford University Press.
Sharma, R. S. (1959). *Political ideas and institutions in ancient India*. Delhi: Motilal Banarasidass.
Sharma, R. S. (2007). *India's ancient past*. New Delhi: Oxford University Press.
Sharp, J. (2006). Corporate social responsibility and development: An anthropological perspective. *Development Southern Africa*, *23*(2), 213–222.
Singh, C. P. (1978). *The Ho tribe of the Singhbhum*. Delhi: Classical Publication.
Singh, K. S. (1971). State formation in tribal society: Some preliminary observations. *Journal of the Indian Anthropological Society*, *6*, 161–181.
Singh, K. S. (1972). Famine, scarcity and economic development in tribal India. In *Tribal situation in India: Proceedings of a seminar* (pp. 388–398). Shimla: Indian Institute of Advanced Studies.
Singh, K. S. (1978). Colonial transformation of tribal society in middle India. *Economic and Political Weekely*, *13*(30), 1221–1232.
Singh, K. S. (1982). Transformation of tribal society: Integration vs assimilation. *Economic and Political Weekly*, *17*(34), 1376–1384.
Singh, K. S. (2002). *Birsa Munda and his movement: A study of millenarian movement, 1874–1901*. Calcutta: Seagull Books.

Singh, K. S. (2006). *Tribal movements in India*. Delhi: Manohar.
Sinha, S. P. (1978). *A report on the Wilkinson's rule in the Kolhan area of Singhbhum district in Bihar*. Ranchi: Bihar Tribal Welfare Research Institute.
Sinha, S. P. (1993). *Conflict and tension in tribal society*. New Delhi: Concept Publishing Company.
Sinn, H.-W. (2010).: *Casino capitalism: How the financial crisis came about and what needs to be done now*. London: Oxford University Press.
Sivaramakrishnan, K. (1996). British imperium and forested zone of anomaly in Bengal, 1767–1833'. *Indian Economic Social History Review*, *33*(3), 243–282.
Sivaramakrishnan, K. (1999). *Modern forests: Statemaking and environmental change in colonial eastern India*. Stanford: Stanford University Press.
Skaria, A. (2001). *Hybrid histories: Forests, frontiers and wildness in western India*. New Delhi: Oxford University Press.
Smelser, N. J. (2013). *Essays in sociological explanation*. Louisiana: Quid Pro Books.
Srivastava, V. K. (2008). Concept of 'tribe' in the draft national tribal policy. *Economic and Political Weekly*, *43*, 29–35.
Steward, J. H. (1972). *Theory of culture change: The methodology of multilinear evolution*. Chicago: University of Illinois Press.
Strange, S. (1986). *Casino capitalism*. New York: Blackwell Publishing.
Streumer, P. (2016). *A land of their own: Samual Richard Tickell and the formation of the autonomous Ho country in Jharkhand 1818–1942*. Houten, the Netherlands: Wakkaman.
Stroshane, T. (1997). The second contradiction of capitalism and Karl Polanyi's the great transformation. *Capitalism Nature Socialism*, *8*(3), 93–116. https://doi.org/10.1080/10455759709358751
Stuligross, D. (2008). Resources, representation, and authority in Jharkhand, India. *Asia Pacific Viewpoint*, *49*(1), 83–97.
Sundar, N. (2005). 'Custom' and 'democracy' in Jharkhand. *Economic and Political Weekely*, *40*(41), 4430–4434.
Sundar, N. (2009a). Framing the political imagination: Custom, democracy, and citizenship. In N. Sundar (Ed.), *Legal grounds: Natural resources, identity, and the law in Jharkhand* (pp. 189–216). New Delhi: Oxford University Press.
Sundar, N. (2009b). Law, policies and practices in Jharkhand. In N. Sundar (Ed.), *Legal grounds: Natural resources, identity, and the law in Jharkhand* (pp. 1–29). New Delhi: Oxford University Press.
Sundar, N. (2015). *Land Acquisition and Modi in Dantewada*. Retrieved, from http://nandinisundar.blogspot.in/2015/05/land-acquisition-and-modi-in-dantewada.html
Susan, P., O'Keefe, P., & Wisner, B. (1983). Global disasters, a radical interpretation. In K. Hewitt (Ed.), *Interpretations of calamity from the viewpoint of human ecology* (pp. 263–283). Boston: Allen and Unwin.
Terborgh, J. (1999). *Requiem for nature*. Washington, DC: Island Press=Shearwater Books.
Tetlock, P. E., Kristel, O. V., & Beth Elson, S. (2000). The psychology of the unthinkable: Taboo trade-offs, forbidden base rates, and heretical counterfactuals. *Journal of Personality and Social Psychology*, *78*(5), 853–870.

Thapar, R. (1999). *From lineage to state: Social formations of the mid-first millennium BC*. Delhi: Oxford University Press.
Thapar, R., & Siddiqieds, M. H. (Eds.). (1979). *Chotanagpur: The pre-colonial and colonial situation in trends in ethnic group relations in Asia and Oceania*. Paris: UNESCO.
Thomas, A. (2000). Introduction. In T. Allen & A. Thomas (Eds.), *Poverty and development in the 1990s* (pp. 1–9). Oxford: Oxford University Press.
Thompson, E. P. (1971). The moral economy of the English crowd in the eighteenth century. *Oxford University Press on Behalf of The Past and Present Society*, *50*(February), 76–136.
Tilly, C., McAdam, D., & Tarrow, S. (2001). *Dynamics of contention*. Cambridge: Cambridge University Press.
Troisi, J. (1978). *Tribal religion: Religion and practice among the santhals*. New Delhi: Manohar.
Tsing, A. (1993). *The realm of the diamond queen*. Princeton, NJ: Princeton University Press.
Upadhya, C. (2009). Law, custom and adivasi identity: Politics of land rights in Chotanagpur. In N. Sundar (Ed.), *Legal Grounds: Natural resources, identity, and the law in Jharkhand* (pp. 30–55). New Delhi: Oxford University Press.
Vaidyanathan, A. (2001). Poverty and development policy. *Economic and Political Weekly*, *38*, 1807–1822.
Vanaik, A. (1990). *The painful transition: Bourgeois democracy in India*. London: Verso.
Vandergeest, P., & Peluso, N. L. (1995). Territorialization and state power in Thailand. *Theory and Society*, *24*(3), 385–426.
Vayda, A. P. (1983). Progressive contextualization: Methods for research in human ecology. *Human Ecology*, *11*(3), 265–281.
Verardo, B. (2003). *Rebels and devotees of Jharkhand: Social, religious and political transformation among the Adivasis of northern India* (Unpublished doctoral dissertation). London School of Economics and Political Science, University of London.
Vidyarthi, L. P., & Rai, B. K. (1977). *The tribal culture of India*. New Delhi: Concept Publishing Company. Retrieved from https://books.google.com/books?id=scgz0Uzn64YC&pgis=1
Vinthagen, S., & Johansson, A. (2013). "Everyday resistance": Exploration of a concept and its theories. *Resistance Studies Magazine*, *1*(1), 1–46. Retrieved from http://rsmag.nfshost.com/wp-content/uploads/Vinthagen-Johansson-2013-Everyday-resistance-Concept-Theory.pdf
Vollard, H. (2009). The logic of political territoriality. *Geopolitics*, *14*(4), 687–706.
Walker, K. J. (1989). The state in environmental management: The ecological dimension. *Political Studies*, *37*(1), 25–38.
Wallerstrein, I. (1974). Dependence in an interdependent world: The limited possibilities of transformation within the capitalist world economy. *African Studies Review*, *17*(1), 1–26.
Walsh, C. (2013). Political and event ecology: Critiques and opportunities for collaboration. *Journal of Political Ecology*, *20*, 199–216. Retrieved from http://jpe.library.arizona.edu/volume_20/PennaFirme.pdf
Walsh, M. (2012). The not-so-great ruaha and hidden histories of an environmental panic in Tanzania. *Journal of Eastern African Studies*, *6*, 303–335.

Watts, M. J. (1989). The agrarian question in Africa: Debating the crisis. *Progress in Human Geography*, *13*(1), 1–41.

Weiner, M. (1978). *Sons of the soil: Migration and ethnic conflict in India*. Delhi: Oxford University Press.

Whitehead, M., Jones, R., & Jones, M. (2007). *The nature of the state: Excavating the political ecologies of the modern state*. Oxford: Oxford University Press.

Wildcat, D., & Pierotti, R. (2000). Finding the indigenous in indigenous studies. *Indigenous Nations Studies Journal*, *1*(1), 61–70.

Williams, R. (1976). *Keywords*. London: Fontana.

Wolf, E. (1972). Ownership and political ecology. *Anthrological Quaterely*, *45*(3), 201–205. https://doi.org/10.2307/3316532

Xaxa, V. (1999). Transformation of tribes in India: Terms of discourse. *Economic & Political Weekly*, *32*(24), 1524–1591.

Xaxa, V. (2008). *State, society, and tribes: Issues in post-colonial India*. Delhi: Pearson Education India.

Xaxa, V. (2011). *Tribes and social exclusion* (No. 2). Retrieved from www.cssscal.org/pdf/unicef/OP_Virginius_Xaxa.pdf

Xaxa, V. (2016). Formation of adivasi/indigenous peoples' identity in India. In M. Radhakrishna (Ed.), *First citizen: Studies on adivasis, tribals and indigenous peoples in India* (pp. 34–52). New Delhi: Oxford University Press.

Yorke, M.P. (1976). *Decisions and analogy. Political structure and discourse among the Ho tribals of India*, PhD Thesis submitted to the School of Oriental and African Studies, University of London. (Unpublished)

Zou, D. V. (2016). Peoples, power and belief in north east India. In M. Radhakrishna (Ed.), *First citizen: Studies on adivasis, tribals and indigenous peoples in India* (pp. 97–125). Delhi: Oxford University Press.

Author Index

Subject Index

For Product Safety Concerns and Information please contact our EU
representative GPSR@taylorandfrancis.com
Taylor & Francis Verlag GmbH, Kaufingerstraße 24, 80331 München, Germany

www.ingramcontent.com/pod-product-compliance
Ingram Content Group UK Ltd.
Pitfield, Milton Keynes, MK11 3LW, UK
UKHW020139200926
475702UK00005B/821
9781032345703